KU-278-966

95 A

8 A

Avril Robarts LRC

Liverpool John Moores University

The FIDIC Digest

Contractual relationships, responsibilities and claims under the fourth edition of the FIDIC Conditions

The FIDIC Digest

Contractual relationships, responsibilities and claims under the fourth edition of the FIDIC Conditions

John G. Sawyer, FRICS, FCIOB, MSE, FCInstArb
C. Arthur Gillott, BEng(Hons), FICE

Thomas Telford, London

Published by Thomas Telford Ltd, Thomas Telford House, 1 Heron Quay, London E14 9XF

First published 1990

British Library Cataloguing in Publication Data
Sawyer, John G.
  The FIDIC digest: Contractual relationships, responsibilities and
  claims under the fourth edition of the FIDIC Conditions
  I. Title II. Gillot, C. A. (Cyril Arthur), *1922–*
  624

ISBN: 0 7277 1510 0

This guide is published on the understanding that the authors are solely responsible for the statements made and opinions expressed in it and that its publication does not necessarily imply that such statements and or opinions are or reflect the views or opinions of the publishers. Every effort has been made to ensure that the statements made and the opinions expressed in this publication provide a safe and accurate guide; however, no liability or responsibility of any kind can be accepted in this respect by the publishers or the authors.

Typeset in Great Britain by MHL Typesetting Limited, Coventry
Printed and bound in Great Britain by

Acknowledgements

Once again we thank our wives for their continuing support, patience and understanding for the time we have devoted to the preparation of this publication rather than to them.

We would also like to thank those readers of our previous Digests on the third edition of the FIDIC Conditions of Contract for their letters telling us of their experiences and of their views on many subjects.

Finally we would like to express our thanks to Mr Kazutake Okuma of Nippon Kokan KK who translated the second edition of our Digest into Japanese and also arranged for its publication in Japan.

John G. Sawyer
C. Arthur Gillott

Preface

This digest is an appraisal of the fourth edition of the FIDIC *Conditions of contract for works of civil engineering construction*, 1987, generally known as 'the Red Book', which is reproduced in appendix 2. It is also concerned with the contractual relationships between the contracting parties and others involved. The publication of the fourth edition, which is no longer designated as international, gives good reason to consider new or changed clauses, and to examine the various duties and obligations of those running the contract. Fortunately there are few major changes from the third edition. Those that have been made generally give greater emphasis to the various obligations of the parties to the contract. There are also innovations to the subjects of claims, insurances, consultation and such like.

Possibly the most notable change is that which brings the employer into greater contact with both the engineer and the contractor over a variety of subjects including, among other things, extra works and extensions of time. The clauses within most construction contracts separate themselves into distinct groups: procedure, authority, responsibility, obligation, risk, change, payments, information, who is who, and what happens when the parties to the contract have a seemingly unresolvable disagreement. Clauses related to most of these matters are examined in particular but those of lesser usage or importance are referred to only in a general way. The fourth edition has incorporated Part III of the third edition into the new Part II, which now deals, among other things, with dredging and reclamation work, but, helpfully, the numbering of the Clauses remains virtually the same in both editions.

Because of the considerable time which will elapse before the third edition is phased out there is little point in referring back to it or making comparisons with the fourth edition. However, a digest on the third edition[1] is available for readers who want to become more familiar with it.

While many clauses have not been fundamentally changed, there are some which have been substantially altered. When applying the FIDIC *Conditions of contract*, or for that matter any legal document, any change of words, however slight, in any clause in a standard form can affect other clauses even if they are not directly related.

The third and fourth editions of the FIDIC *Conditions of contract* are both intended for Works of Civil Engineering Construction which are to be measured and evaluated by Bills of Quantities. They are not suitable, without alteration, for works using a lump sum, fixed price, or target-cost contract where the basis for payment is not admeasurement, nor in their present form for design and build work.

In both sets of conditions the order of importance concerning any construction project is as follows

(a) Principals or Parties to the Contract: Employer; Contractor
(b) supported by: the Engineer, his Representatives and Assistants; Nominated and other Sub-Contractors and Suppliers
(c) others involved: Insurers; Providers of Geological details etc; Designers and Specialist Engineers
(d) by invitation: Arbitrator or Arbitrators.

It is the Employer and not the Engineer or Contractor in this particular form of contract who is responsible for providing or arranging the funds which enable the contract itself to come into being; it is the Engineer and the Contractor together who provide the technical skills necessary for the works to be constructed. If they all perform their various obligations and duties properly and on time, few problems will arise, but if anyone neglects to honour the commitments he has undertaken, then the use of the Clauses comprising the Conditions becomes essential, and disputes will arise which will have to be resolved.

Contents

Introduction

The publication of the fourth edition of the FIDIC *Conditions of contract* in no way detracts from the necessity of understanding the manner by which its Conditions are enforced. This is established through the Laws of the Land which govern the Contract and it is therefore of primary importance that the identification of the law applicable to the Contract is made clear, as well as the ruling language where the Contract documents are written in more than one language.

So important is the identification of the language the Conditions require that a definition needs to be made as to which the ruling language is where two or more languages are used within the Contract. The establishment of the ruling language determines that in any interpretation of the Conditions a second or other language is for convenience only and not to be considered authoritative if differences of translation arise. When a document is prepared in one language and translated into another it is a good idea to have the second language version independently translated back into the original language as a check, because, notwithstanding the efficiency of the translators, a number of identifiable differences are often discovered.

Sub-Clause 5.1 and Part II Sub-Clause 5.1(a)

A word of caution about language would not go amiss here. When dealing with international business in foreign languages, one must always be sure that the idea or concept perceived by each Party for a given word is the same for all. Because some people of different countries speak English very well, one must never assume that their understanding of the English words and grammar is necessarily correct — the same can apply when two people using their own language are communicating with each other.

Because of the many variations in legal philosophies throughout the world, the need cannot be overemphasised for all Parties involved to have an adequate understanding of the particular legal system which governs the Contract. They should also be aware of the importance of receiving professional guidance on Statutes, Ordinances, National or Inter-State legal requirements which can override or influence the particular clauses contained within FIDIC, or indeed any other Conditions of Contract.

Sub-Clause 5.1(b)

When using the FIDIC *Conditions of contract* there are many Parties who prefer to nominate the established laws of England to govern its interpretation, but the legal system of the country where any dispute is resolved, either through the courts or arbitration, may require the application of its own laws and would therefore negate such choice.

The attitudes of courts may vary in the interpretation of contracts. In some countries the courts will enforce a literal interpretation if this leads to a clear and unambiguous result, while in other countries the courts will look more to the intentions of the Parties. The applicable law will also determine the extent to which the Parties are bound to carry out their contractual obligations. Under English law there is no concept of 'force majeure' unless specifically incorporated within the Contract — only the very restricted concept of 'frustration', where a supervening event prevents performance by one or both of the Parties. Under systems based on the Code Napoleon the Parties to a Contract are relieved of their obligations to the extent that they are prevented from performing them by a case of force majeure. Courts under such systems of law may, where exceptional circumstances render the performance of one Party's obligations so onerous as to incur heavy loss (but without rendering performance impossible), reduce the onerous obligations to a reasonable level of equity so required. This potential conflict between force majeure and frustration has been prevented in the fourth edition by Clause 66.1 Release from Performance.

Once the importance of understanding the legal system which governs the Contract is recognised, it is equally important for those involved with the Project to understand in practical terms the nature of their responsibilities and liabilities, and the mode of conduct expected of them. Everybody involved with the Contract has an important part to play in ensuring the successful completion of the Project, and this means the complete fulfilment of all their obligations to each other.

The FIDIC document demands a system of communication which cannot be shortened or ignored without putting at risk the rights of any one Party to the Contract. It is therefore important to remember that when communications are required to be in writing, this means precisely that, and in like manner, the giving of notices, instructions or certificates should follow the timing and procedures precisely as dictated by the Conditions of Contract, and in the detail required. In the event of a dispute, the legal representatives of all Parties will rely more than anything else on written evidence, documents, and diaries, and if one Party has failed to fulfil its obligations properly in this respect, then its chances of success may be considerably reduced.

Those who are familiar with the fourth and fifth editions of the Conditions of Contract issued by the British Institution of Civil Engineers (ICE) will be well aware of points of similarity contained in Clauses in both of these ICE editions and in the FIDIC Conditions. They should not, however, because of their knowledge of the ICE Conditions, assume that FIDIC is equally applicable in usage — it is not. FIDIC is a form of Contract not necessarily subject to the laws of England; nor is the wording precisely the same as the ICE Conditions either in syntax or definition. The FIDIC Contract must therefore be regarded separately and in its own right, and should be read and fully understood with these points in mind. The third edition of the FIDIC Conditions has been translated into various languages such as German and Spanish.

1. Contractual relationships

Considerations and relationships

For a Project to be created and before a Contract can be signed with the successful tendering Contractor, the Employer has to examine what he wants to achieve and make a number of decisions. Which responsibilities will he accept for himself and which will he delegate to others skilled in doing things which he is not competent to do? A great deal of work has to be done by the Employer and his appointed Engineer to establish the viability of the Project. Then the engineering design has to be undertaken, all powers and legal rights secured and proper Contract documents prepared.

Everyone involved with the Project knows that the intentions of the Parties are deemed to be expressed in the Contract documents but that some of these intentions are stated explicitly whereas others are implied. Both types need to be recognised. All should be well if both Parties to the Contract, together with the Engineer, the Sub-Contractors and the Suppliers, behave in accordance with the documents. Trouble starts when any of the above-named behave incorrectly and not in accordance with the Contract.

To understand what is expected of those participating in the Project their involvement must be looked at in detail.

The Employer

There should be no doubt as to who the Employer actually is in that he is named and his address given in Part II. He can be changed during the life of the Contract but if the Employer wishes to assign the Contract, this can only be done with the Contractor's consent.

Clause 1.1(a)(i)

There are various tasks the Employer is obliged to do.

First task

He has to develop the concept of the Project, examine its feasibility and financial viability and prepare a cost plan. He must have determination, authority, finance and energy to be able to proceed, and not only to design and build, but also to occupy, operate, maintain and support the Project throughout its life cycle. It is also important that he obtains, or is already in possession of, adequate technical and commercial resources to fulfil the purpose of the entire Project.

3

Second task

He must secure the political and legal support of his Government to be able to obtain all necessary decrees, statutes, laws or regulations for construction, operation, maintenance and ultimate renewal or disposal of the Project. He must also be able to obtain adequate supplies of energy, water, raw materials, labour, staff, communications and markets, and be able to dispose of any waste products. He should ensure, if he commits himself to help the Contractor import and export the necessary parts or plant for the Project, that he is in a position to do so.

Third task

A design brief has to be established and a design team appointed to develop the size, shape, location and functional requirements of the Project and then to prepare positive plans for any procurement, its construction and operation.

Fourth task

A supervising Engineer for the construction phase has to be appointed, whether it is the leader of the original design team or another Engineer. The appointed Engineer, who will be named in the Contract, will then be jointly involved with the Employer in determining the details of the Conditions of Contract to be used, the pre-qualification phase of the selection of the Contractor and the Tender Documents to be used. His appointment is made in a separate Contract with the Employer. The details of this are not disclosed to the Contractor, but the Employer is obliged to disclose any matter for which he requires to give approval before it can be implemented by the Engineer.

Sub-Clause 1.1(a)(iv)

Sub-Clause 2.1(b)

Many of the details to be considered will include giving a definition of the law and language of the Contract, the currency to be used, rates of exchange, financial plans, bonds to be required, (bid bonds, advance payment bonds and the traditional performance bonds) insurances, bonus payments, penalties or liquidated damages, timing for completion or sectional completion (including a programme for pre-qualification, tendering and appointment of the Contractor), Employer plant and material supply and dates of delivery, nomination of Suppliers and specialist Sub-Contractors.

The Employer is also responsible for the provision of facilities, the definition and availability of the Site for the Works, access to the Site, and for relations with other affected parties. He must also define the Taking Over of Works, the Defects Liability Period, the condition for commissioning, testing and putting into operation of the Works, and many other matters.

Furthermore the Employer must ensure that adequate funds are allocated and usefully spent on the preliminary studies, site investigations, soil surveys, precise surveys, acquisition of land for the Works and Temporary Works. The budget for construction and commissioning and testing must also be adequate. A proper cash flow forecast by the Employer is vital as the lack of provision for advance and mobilisation payments has jeopardised many projects in the past.

A particularly important decision to be made by the Employer is how to deal with Insurances. While Part I deals with this subject in considerable detail, and it is the responsibility of the Contractor, in Part II the Employer has the option of handling Insurances himself where a number of separate Contractors are employed on a single project or where a phased Take-Over is involved. A vital matter often neglected is the security and provision of essential records and documents, many of which have to be maintained for ten, fifteen, twenty years or more. The Employer should determine the policy, procedures and practice to be adopted for the safe keeping of all site records.

Clauses 20, 21, 22, 23, and 25

Fifth task

A decision must be made about which form of Contract is to be used. If the work is mainly civil engineering and is to be valued by admeasurement based on a Bill of Quantities then the FIDIC 'Red Book' (see appendix 2) standard form will be suitable. These conditions are not intended to cover works valued on a Lump Sum, Cost Plus or Target Cost of a Design and Build basis. It is not recommended that the FIDIC 'Red Book' be varied to accommodate these other works as it is dangerous to amend any form of Contract to cover work for which it was not intended. FIDIC has published a tendering procedure[2] for obtaining and evaluating Tenders for civil engineering Contracts and this is recommended reading for Employers, Engineers and Contractors.

The first, second and third editions of the FIDIC *Conditions of contract for works of civil engineering construction* did not give the Employer many duties but the fourth edition allocates certain matters to him more specifically. Clause 2.1 in Part II lists some of the Engineer's duties for which the specific approval of the Employer must be obtained before they are carried out.

There are twenty-three occasions when the Engineer must consult the Employer and the Contractor and only one occasion when the Employer consults the Contractor on a matter of settlement. The Employer is the more important Party to the Contract because it is his Project and his money which provide employment for the Engineer and the Contractor.

The Engineer

The Engineer is not a party to the Contract between the Employer and the Contractor. He is named in the Contract and has a number of duties and responsibilities requiring his expertise in technical design and management. His duties, as required for the administration of the Contract between the Employer and the Contractor, are set out clearly in the Conditions.

The Engineer has a separate agreement between himself and the Employer which covers all the work he is required to carry out for the Employer. This may embrace some or all of the following: feasibility studies, design of the Works, preparation of the Tender Documents, assessment of Tenders, supervision of the Works, supervision of commissioning and testing, supervision of operation of the Works, training of personnel, preparation of maintenance and operating manuals, and suchlike.

The agreement between the Employer and Engineer might be based on the British Association of Consulting Engineers' model form of Agreement or on the appropriate FIDIC model form, several versions of which have been published.

Pre-Contract

During the pre-Contract period the Engineer becomes involved with a number of activities necessary for the Contract documents to be issued. Some of these are of a design and technical nature whereas others deal with matters necessary to obtain competitive Tenders from selected Contractors. These include working with the Employer to develop the design brief, investigate and identify all the resources needed to undertake the total Project, examining the financial risks involved especially those that are unforeseen or special, and preparing realistic financial and operational plans. The Engineer must ensure that the Employer accepts that certain decisions and actions are required from him to suit the programme, to meet various target dates and most particularly to ensure the availability of sufficient funds to honour the Engineer's Certificates for Payment as required by the Contract.

The Employer should also confirm that the acquisition and availability of the Site and means of access have been organised, and that any governmental decrees and planning permissions have all received his attention so that the Engineer can plan a time schedule to form the basis of the Contract. The Engineer also needs to know a suitable Commencement Date and whether the Employer wants the works to be completed in sections or as a whole. These matters will be identified in the appendix (Part II of the Conditions of Contract).

The detailed design of the Works themselves and of any Temporary Works which the Engineer undertakes must be dealt with. Clear Working Drawings, specifications and Bills of Quantities, all of which require detailed preparation, must also be provided. All other essential details should be agreed between the Engineer and Employer, and also how the Tendering Contractors are to be invited to tender, whether by open invitation or by pre-qualification. A time schedule as to when all this should happen also has to be agreed.

The Engineer would be wise to inform the Employer and his staff about the role of the Engineer under the FIDIC Conditions and to emphasise its importance in ensuring fair dealings between the Contractor and the Employer.

Tender Procedure

The Engineer must make certain that all Tenderers receive the same Tender information, and have equal opportunities to visit the Site and examine all the available information relevant to the Works. The Engineer should supervise the receipt, security and opening of the Tenders at the appointed time.

Once the Tenders are open the Engineer must study each one carefully, correct errors, examine method statements and programmes if submitted with the Tender, and pay particular regard to any qualifications that could invalidate the Contractor's offer. The Engineer must establish that the several Tender sums are based on the same conditions for each Tenderer so that a true comparison can be made and a shortlist made up for the Employer to study. The Engineer should also check the statements made about the resources and past experience of the Contractors in the lower range of Tenders to ensure that the successful Tenderer is capable of carrying out the Works and financially competent to complete them.

One of the final duties of the Engineer in the pre-Tender period is to prepare a report for the Employer setting out the merits of the various bids and to recommend, with reasons, the most suitable Tenderer he believes will best serve the interests of the Employer. The Employer, however, must make his own decision.

Post-Contract

When the formalities of appointing the successful Tenderer are completed and the Works commenced the Engineer has certain matters to attend to. Both he and his Representative should keep diaries of their involvement with the Contract. These should include important matters which might need to be referred back to. In particular, the Engineer's Representative should record daily all matters of site interest such as weather conditions, progress, and the Contractor's plant and labour on site, whether working or otherwise. Progress photographs would improve the standard of record keeping. These should be taken at regular predetermined intervals and on special occasions, for example after storms, earthquakes, floods or strikes. There should be regular site meetings and the preparation and agreement of the minutes of such meetings should follow as soon as possible, preferably the day after. The Engineer himself should make regular visits to the Site to be kept fully updated as to progress. This provides the opportunity for both his Representative and the Contractor to discuss with him any matters of importance to the Contract.

It has been proven time and again that the Engineer, more than anyone else with a key role in the Project, can ensure for the Employer that the Works are completed on time, that they are of the required quality, and also that the Contractor gets a fair and proper payment and is given the opportunity to make a profit.

At this stage it is worth a glance at some Clauses in the FIDIC model form of agreement between the Employer and the Engineer for the Project. Clauses 2.3.1 and 2.3.5 are as follows.

Clause 2.3.1

The Consulting Engineer shall exercise all reasonable skill, care and diligence in the performance of the Services under the Agreement and shall carry out all his responsibilities in accordance with recognised professional standards.

The Consulting Engineer shall in all professional matters act as a faithful adviser to the Client and, in so far as any of his duties are discretionary, act fairly as between the Client and third parties.

The Consulting Engineer, his employees and sub-contractors, whilst in the country in which the Works are to be carried out, shall respect the laws and customs of that country.

Clause 2.3.5

The Consulting Engineer, when in charge of the supervision of Works under construction, shall have authority to make minor alterations to design as may be necessary or expedient but he shall obtain the prior approval of the Client to any substantial modification of the design and costs of the said Works and to any instruction to a Contractor which constitutes a major variation, omission or addition to the latter's contract. In the event of any emergency, however, which in the opinion of the Consulting Engineer requires immediate action in the Client's interest the Consulting Engineer shall have authority to issue such orders as required on behalf of and at the expense of the Client. The Consulting Engineer must inform the Client immediately of any orders issued without prior consent which will result in additional cost to the Client and follow up such advice as soon as possible with an estimate of the probable cost.

Clause 2.3.5 restricts the actions of the Engineer in making alterations to designs and varying the Works. It is quite proper that the Engineer must seek approval from his client before making major changes, but any matters for which the Engineer has to seek the Employer's approval before dealing with them must be notified to the Contractor as such within Part II of the Conditions of Contract. Nevertheless, it must be up to the Engineer and the Employer alike to make decisions on any changes quickly, so that the Contractor knows exactly where he stands for payment and any extensions of time.

The Contractor (pre-Tender)

Before contemplating tendering for Works anywhere in the world the Contractor must be certain that he can provide the resources necessary to undertake and complete the Works relevant to a potential Employer's Project. Furthermore he must have enough financial resources to enable him to wait for the Final Settlement and payment in full of the Contract Price as he knows he must. He needs lots of stamina and faith in his own abilities as well as those of his supporters.

He is required to provide such resources as finance, the ability to trade internationally, staff with the necessary technical, commercial and management abilities, skilled workmen and supervisors, materials, plant and equipment and a wealth of experience from all involved. Although he might not have all these resources within his own organisation he must know how, where and when he can acquire the resources he does not possess.

Possibly one of the most important considerations for the Contractor is the risk that the money he spends on tendering for the Project will never be recovered if he does not secure the Contract. The cost and wasted effort can be considerable. He must have the skill to price work competitively but without taking too much off the price and making false economies to win the Contract. If he errs by making too many price cuts in tendering he could secure a Contract at the cost of leaving too large a margin between himself and the next highest bidder. It is also an advantage for the Contractor to have a reasonable standing in the industry so he can procure the various bonds required and purchase the necessary Insurances at competitive rates.

The Contractor must recognise that by submitting a Tender for the Works of the Project he warrants that the Works are capable of being physically constructed by his organisation.

Once the Contract is secured, the Contractor is obliged to execute the construction of the Works with due care, expedition, and without delay, to provide all things necessary for these purposes and be responsible for the stability and safety of all Site operations. Certain exceptions can exist but these must be expressly stated in the Contract documents.

It must be clearly understood that the Contractor is required to complete the Contract on time, subject to any adjustments permitted under the Contract, and to fulfil all his obligations just as he expects the Employer to fulfil his. He must follow all the procedures as laid down in the Conditions of Contract, and during the course of construction carry out the instructions of the Engineer, and make certain that he, the Contractor, gives and receives all notices to and from the right people at the right time and that they are sent to the correct addresses, which are set out in Part II. Most important of all is that he complies with the requirements of the Contract, the law of the country in which the Project is situated, and the laws of the land by which his own business is governed.

2.   Schedules of responsibilities

Both Parties to the Contract, and the Engineer, have a number of responsibilities which are set out very clearly in the Conditions of Contract. These must be fully observed if work is to progress properly and without too many arguments. Some Clauses require particular attention by each of the three main participants of the Contract. So that they can be identified without difficulty, these have been listed under the name of each of those responsible — the Employer, the Contractor and the Engineer.

In each section the Clause number and description are given in the left-hand column and the corresponding responsibility is given in the right-hand column.

Responsibility of Employer

1.1.(a)(i) Definition of Employer	To give precise definition in Sub-Clause 1.1 in Part II; to assign later if necessary after seeking the Contractor's agreement.
1.1.(a)(iii) Definition of Sub-Contractor	To name in Contract documents any person appointed as Sub-Contractor for part of the Works.
1.1.(a)(iv) Definition of Engineer	To appoint the Engineer before inviting Tenders and enter details in Sub-Clause 1.1 Part II.
1.1.(f)(vii) Definition of Site	To define the Site and other places required for Works and include details in the Contract.
1.5. Notices, Consents, Approvals, Certificates and Determinations	To ensure all notices etc. which are his responsibilities are in writing and not unreasonably withheld or delayed.
2.1. Engineer's Duties and Authority	To define the duties of the Engineer which require prior or special approval by the Employer and state them in Part II Sub-Clause 2.1.
3.1. Assignment of Contract	To give consent or otherwise to any assignment of Contract or part thereof by the Contractor.
4.2. Assignment of Sub-Contractors' Obligations	To request and pay for the benefit of the Sub-Contractors' obligations assigned to the Employer by the Contractor for periods after the end of the Defects Liability Period.
6.4. Delays and Cost of Delay of Drawings	To be consulted, along with the Contractor, by the Engineer to enable him to determine extensions of time and costs to be added to Contract Price.

9.1. Contract Agreement	To request the Contractor to enter into and execute the Contract Agreement prepared by the Employer at his cost.
10.1. Performance Security	To approve any institution providing security and the form of security to be provided by the Contractor.
10.3. Claims under Performance Security	To notify the Contractor before making a claim and to state the nature of the default.
11.1. Inspection of Site	To make available at the time of Tender relevant data appertaining to hydrological and sub-surface conditions. See Part II Sub-Clause 11.1 for Dredging and Reclamation Work
12.2. Adverse Physical Obstructions or Conditions	To be consulted, along with the Contractor, by the Engineer to enable him to determine extensions of time and costs to be added to the Contract Price.
19.2. Employer's Responsibilities	When carrying out work on Site or employing other Contractors (Clause 31), to be responsible for safety of all persons and to keep the Site in an orderly state to avoid danger.
20.1. Care of Works	To take care of the Works, Sections or Parts after the date of the relevant Taking-Over Certificate.
20.3. Loss or Damage Due to Employer's Risks	To be liable for all or part of loss or damage arising from any of the Employer's Risks.
20.4. Employer's Risks	To accept liability for loss or damage arising from any risks listed in this Sub-Clause.
21.1. Insurance of Works and Contractor's Equipment	To be a joint name in the Contractor's Insurance of the Works or, as set out in Part II Clauses 21, 23 and 25, to arrange Insurance of the Works and give details of such in Tender Documents.
21.3. Responsibility for Amounts not Recovered	To bear amounts not insured or not recovered from insurers in accordance with risks set out in Clause 20.
22.1. Damage to Persons and Property	To be indemnified by the Contractor other than for exceptions listed in Sub-Clause 22.2.
22.3. Indemnity by Employer	To indemnify the Contractor against claims, proceedings, damages, costs, charges and expenses arising from exceptions defined in Sub-Clause 22.2.
23.1. Third Party Insurance (including Employer's Property)	To be insured jointly with the Contractor against third party liabilities.
23.3. Cross Liabilities	To be considered separately insured by a cross-liability Clause in insurance defined in Sub-Clause 23.1.
24.1. Accident or Injury to Workmen	The Employer is not liable for death or injuries to employees of the Contractor and Sub-Contractors except when caused by his own act or default; to be indemnified by the Contractor.
24.2. Insurance Against Accident to Workmen	To be indemnified under the policies of the Sub-Contractors but to be given evidence of such when required.
25.1. Evidence and Terms of Insurances	To receive evidence and sight of insurance policies from the Contractor before start of Work and within 84 days

of the Commencement Date respectively; to approve terms of insurance by the Contractor.

25.2. Adequacy of Insurances	To ask the Contractor to provide policies in force and receipts for payment as necessary.
25.3. Remedy on Contractor's Failure to Insure	The Employer may effect and keep in force Insurances, pay premiums and recover money from the Contractor if he fails to effect and keep in force insurances required under the Contract.
25.4. Compliance with Policy Conditions	To indemnify the Contractor against losses or claims arising if the Employer fails to comply with conditions imposed by insurance policies effected pursuant to the Contract.
26.1. Compliance with Statutes, Regulations	To be indemnified by the Contractor against all penalties and liability of breach of provisions of statutes, regulations, laws etc., but to be responsible for obtaining planning, zoning or similar permissions for Works to proceed and to indemnify the Contractor in accordance with Sub-Clause 22.3.
27.1. Fossils	To accept ownership as between the Contractor and Employer of all fossils, coins and articles of value discovered on the Site; to be consulted, along with the Contractor, by the Engineer for him to determine extensions in time and costs to be added to Contract Price.
28.1. Patent Rights	To be indemnified by the Contractor against claims.
29.1. Interference with Traffic and Adjoining Properties	To be indemnified by the Contractor against claims.
30.2. Transport of Contractor's Equipment or Temporary Works	To be indemnified by the Contractor against damages to roads and bridges.
30.3. Transport of Materials or Plant	To negotiate and settle certain claims arising from damage to roads and bridges and to keep the Contractor indemnified; to be consulted, along with the Contractor, by the Engineer if the Contractor is deemed to be responsible for causing damage. The Engineer determines the amount due to the Employer from the Contractor. To notify the Contractor and consult him before settling any claims with authorities.
30.4. Waterborne Traffic	As above if traffic is waterborne.
36.5. Engineer's Determination where Tests not Provided for	To be consulted, along with the Contractor, by the Engineer for him to determine extension of time or costs to be added to the Contract Price for tests instructed by the Engineer not otherwise provided for.
37.4. Rejection	To be consulted, along with the Contractor, by the Engineer for him to determine sums due to the Employer from the Contractor for costs incurred by the Employer by repetition of tests.
38.2. Uncovering and Making Openings	To be consulted, along with the Contractor, by the Engineer for him to determine the cost to be added to the Contract Price in respect of uncovering satisfactory work.

39.2. Default of Contractor in Compliance	To employ and pay others to carry out the orders of the Engineer if the Contractor fails to remedy improper work etc., and to be consulted, along with the Contractor, by the Engineer for him to determine the amount due to the Employer from the Contractor.
40.2. Engineer's Determination following Suspension	To be consulted, along with the Contractor, by the Engineer for him to determine extension of time and cost to be added to the Contract price, arising from suspension of Works through reasons beyond the Contractor's control.
40.3. Suspension lasting more than 84 Days	To be aware that the Contractor may omit works and terminate his employment under Sub-Clause 69.1.
42.1. Possession of Site and Access Thereto	To give possession of the Site and access thereto as required by the Contractor in accordance with agreed programmes, proposals or Contract requirements.
42.2. Failure to Give Possession	To be consulted, along with the Contractor, by the Engineer for him to determine extension of time and costs to be added to the Contract Price when Site and access are not given to the Contractor as required.
44.1. Extension of Time for Completion	To be consulted, along with the Contractor, by the Engineer for him to determine extensions in time due to the Contractor.
46.1. Rate of Progress	To be consulted, along with the Contractor, by the Engineer for him to determine costs due to the Employer from the Contractor for additional supervisory costs.
47.1. Liquidated Damages for Delay	To recover money from the Contractor when damages are due.
47.3. (Part II) Bonus for Completion	To decide at pre-Tender stage whether to introduce a bonus scheme for early completion, and to introduce such a scheme.
48.2. Taking-Over of Sections or Parts	To occupy or use Sections or Parts of the Works whether provided for in the Contract or not, but to be aware that the Engineer will need to issue a Taking-Over Certificate to suit if requested to do so by the Contractor.
49.4. Contractor's Failure to Carry Out Instructions	To employ and pay others to carry out instructions to remedy defects if the Contractor defaults; to be consulted, along with the Contractor, by the Engineer for him to determine monies due to the Employer by the Contractor arising from such default.
50.1. Contractor to Search	To be consulted, along with the Contractor, by the Engineer for him to determine costs incurred by the Contractor in searching to be added to the Contract Price.
51.1. Variations	The Employer cannot carry out work omitted by the Engineer.
52.1. Valuation of Variations; 52.2. Power of Engineer to Fix Rates	To be consulted, along with the Contractor, by the Engineer, for him to agree suitable rates and prices with the Contractor.

52.3. Variations Exceeding 15 per cent	To be consulted, along with the Contractor, by the Engineer for him to agree with the Contractor such further sums to be added to or deducted from the Contract Price.
53.5. Payment of Claims	To be consulted, along with the Contractor, by the Engineer for him to certify sums due to the Contractor for claims.
54.3. Customers Clearance; 54.4. Re-export of Contractor's Equipment	To assist Contractor to clear customs when importing plant and to obtain government consent when re-exporting it.
54.5. Conditions of Hire of Contractor's Equipment	To be entitled to use hired equipment brought to the Site by the Contractor in the event of the Contractor's default under Clause 63. Under Sub-Clause 54.6, the costs of re-hire are to be part of the cost of executing and completing the Works recoverable from the Contractor.
59.1. Definition of 'Nominated Sub-Contractors'	The Employer can nominate Sub-Contractors for Parts of the Works as long as the Contractor accepts the nominations.
59.5. Certification of Payments to Nominated Sub-Contractors	To be entitled to pay the nominated Sub-Contractors directly on the Engineer's certificate if the Contractor fails to pay.
60.2. Monthly Payments	To receive the certificate from the Engineer for the amount of monthly payment due to the Contractor.
60.7. Discharge	To receive written discharge from the Contractor confirming that the Final Statement submitted by him represents the full and final settlement.
60.8. Final Certificate	To receive from the Engineer a Final Certificate 28 days after the Contractor has submitted his Final Statement and Discharge.
60.9. Cessation of Employer's Liability	The Employer is not liable to the Contractor for any matter or thing arising unless the Contractor has included a claim for it in his Final Statement and in his Statement at Completion (Sub-Clause 60.5).
60.10. Time for Payment	To pay the Contractor the amount certified by the Engineer for interim payments within 28 days of receipt of the interim certificate; to pay the Contractor the amount due under the Engineer's Final Certificate within 56 days of receipt of the Final Certificate; to pay the Contractor interest at the rate stated in the Appendix on all sums unpaid after the expiry of these periods. The Employer is liable to have the Works suspended or slowed down by the Contractor under Clause 69 if payments are not made within the times stated.
62.1. Defects Liability Certificate	To receive from the Engineer a Defects Liability Certificate when the Contractor has completed all his obligations under the Contract.
62.2. Unfulfilled Obligations	To be liable to fulfil any obligations incurred prior to the issue of the Defects Liability Certificate which remain unperformed at the time such a certificate was issued.
63.1. Default of Contractor	In the event of the Contractor's failure, the Employer can

enter the Site and the Works, terminate the Contractor's employment and complete the Works.

63.3. Payment after Termination

The Employer is not liable to pay the Contractor any further amounts until the expiry of the Defects Liability Period and until the Engineer has ascertained and certified any amounts due to the Contractor after allowing for all costs incurred by the Employer in completing the Works. The Employer can recover any sums due from the Contractor as certified by the Engineer.

63.4. Assignment of Benefit of Agreement

To receive from the Contractor, subject to the Engineer's instructions, the benefits of any agreement to supply goods, materials or services, or to execute work entered into by the Contractor for the purposes of the Contract.

64.1. Urgent Remedial Work

The Employer may employ and pay others to undertake urgent repairs if the Contractor is unable or unwilling to do such works; to be consulted, along with the Contractor, by the Engineer for him to determine costs incurred by the Employer recoverable from the Contractor.

65.5. Increased Costs arising from Special Risks

To be consulted, along with the Contractor, by the Engineer for him to determine costs to be added to the Contract Price arising from Special Risks.

65.6. Outbreak of War

The Employer is entitled to terminate the Contract any time after the outbreak of war.

65.8. Payment if Contract Terminated

To be consulted, along with the Contractor, by the Engineer for him to determine any sums due to the Contractor under this Sub-Clause.

66.1. Payment in Event of Release from Performance

To pay the Contractor as if the Contract had been terminated under Clause 65.

67.1. Engineer's Decision

To refer disputes with the Contractor to the Engineer for a decision under Clause 67; to follow procedures laid down.

67.2. Amicable Settlement

To attempt to settle disputes amicably within 56 days of intention to commence Arbitration unless otherwise agreed before actually commencing Arbitration.

67.3. Arbitration

To refer dispute for settlement under Rules of Conciliation and Arbitration of the International Chamber of Commerce (ICC), or to use other dispute procedures set out in Part II Sub-Clause 67.3.

67.4. Failure to Comply with Engineer's Decision

May refer to Arbitration under Sub-Clause 67.3 any dispute arising because the Contractor fails to comply with an earlier final and binding decision of the Engineer.

68.1. Notice to Contractor; 68.2. Notice to Employer and Engineer; 68.3. Change of Address

To follow procedure laid down for serving and receiving of notices set out in Clause 68.

69.1. Default of Employer

To receive notice from the Contractor to terminate his employment in the event of actions of the Employer set out in this Sub-Clause.

69.3. Payment on Termination

To be under the same obligations to the Contractor as if

the Contract was terminated under Clause 65 but to pay in addition loss or damage incurred by the Contractor as a result of the termination under Clause 69.

69.4. Contractor's Entitlement to Suspend Work	To be consulted, along with the Contractor, by the Engineer for him to determine extension of time and costs to be added to the Contract Price in the event of late payments and when the Contractor elects to suspend work or reduce the rate of work.
70.2. Subsequent Legislation	To be consulted, along with the Contractor, by the Engineer for him to determine costs to be added to or deducted from the Contract Price arising from changes in Statutes, Ordinances, Decrees, or Laws made after a date 28 days prior to the date for submission of Tenders.
71.1. Currency Restrictions, and 72.1. Rates of Exchange	To determine policy covering those matters to be included in Tender Documents and to set out procedures and principles to be followed in Part II.
11.1, 12.2, 18.1, 19.1, 28.2, 40.1, 40.2, 45.1, 49.5, 50.2, 51.1. Dredging and Reclamation Work, various Clauses	To make appropriate changes to documents as indicated in various related Clauses set out in Part II.

Responsibililty of Engineer

1.5. Notices, Consents, Approvals, Certificates and Determinations	To ensure all notices etc. are in writing and not unreasonably withheld or delayed.
2.1. Engineer's Duties and Authority	(a) To carry out duties specified in the Contract. (b) To exercise authority specified or implied by the Contract but to obtain the specific approval of the Employer for authorities listed in Part II Sub-Clause 2.1. (c) The Engineer cannot relieve the Contractor of his obligations under the Contract unless expressly stated therein.
2.2. Engineer's Representative	To appoint the Engineer's Representative and be responsible for his actions.
2.3. Engineer's Authority to Delegate	To delegate to the Engineer's Representative any of the duties and authorities vested in the Engineer by the Employer and by the Contract. Any delegation or revocation of the delegation must be in writing and can be effective only when copies of the delegation have been delivered to the Employer and the Contractor. The Engineer can vary or reverse the decisions of the Engineer's Representative.
2.4. Appointment of Assistants	To appoint or to get the Engineer's Representative to appoint persons to assist him in carrying out his duties (Sub-Clause 2.2); to notify the Contractor of names, duties and authorities.
2.5. Instructions in Writing	To give instructions in writing and to confirm any oral instructions in writing. The Contractor may confirm in

	writing any oral instructions from the Engineer. This Sub-Clause applies to the Engineer's Representative and his assistants.
2.6. Engineer to Act Impartially	To exercise his discretion impartially when giving decisions etc. expressing approval, determining value or otherwise taking action affecting the Employer and the Contractor. (All subject to application of Clause 67 Settlement of Disputes.)
4.1. Sub-Contracting	To give consent in writing to the Contractor to sub-let part of the Works.
5.2. Priority of Contract Documents	To explain and adjust ambiguities and discrepancies in the Contract documents and instruct the Contractor thereon.
6.1. Custody and Supply of Drawings and Documents	To retain custody of Drawings; to provide two copies free of charge to the Contractor; to give consent to the Contractor to show Drawings, Specifications etc. to third parties if strictly necessary for the purposes of the Contract; to approve Drawings, Specifications and other documents provided by the Contractor when he undertakes design work under Clause 7; to request in writing further copies of Drawings, Specifications etc. from the Contractor for use by the Employer.
6.2. One Copy of Drawings to be Kept on Site	To use Drawings kept on Site by the Contractor and authorise in writing use of such by other persons.
6.3. Disruption of Progress	To receive notice from the Contractor when further drawings or instructions are required from the Engineer to avoid delay or disruption to the Works.
6.4. Delays and Cost of Delay of Drawings	If the Engineer fails to issue Drawings or instructions requested by the Contractor under Sub-Clause 6.3 within reasonable time, he must consult the Employer and Contractor and determine any extension of time to the Contractor under Clause 44; costs to be added to the Contract Price and the Contractor and Employer to be notified accordingly.
6.5. Failure by Contractor to Submit Drawings	To take into account in determination under Sub-Clause 6.4 any failure on the part of the Contractor to submit Drawings, Specifications etc. which in turn causes the Engineer to fail to provide Drawings and instructions.
7.1. Supplementary Drawings and Instructions	The Engineer has authority to issue to the Contractor supplementary Drawings and instructions as necessary.
7.2. Permanent Works Designed by Contractor	To receive for approval Drawings, Specifications, calculations and other information of Permanent Works designed by the Contractor as required by the Contract; to approve operations and maintenance manuals and drawings on completion submitted by the Contractor before taking over in accordance with Clause 48.
7.3. Responsibility Unaffected by Approval	Approval given under the above Sub-Clauses does not relieve the Contractor of his responsibilities.

12.2. Adverse Physical Obstructions or Conditions	To receive notice from the Contractor when encountering non-foreseeable obstructions or conditions on Site. If in his opinion such obstructions etc. were not reasonably foreseen, the Engineer will, on receipt of notice, consult the Employer and the Contractor and determine any extension of time and costs to be added to the Contract Price and notify the Contractor and the Employer. The Engineer may instruct the Contractor to take measures to overcome obstructions etc., or to accept measures taken by the Contractor.
13.1. Work to be in Accordance with Contract	The Engineer has to be satisfied with the Contractor's execution and completion of the Works including remedying defects in accordance with the Contract. The Engineer is sole conveyor of instructions to the Contractor.
14.1. Programme to be Submitted	To receive from the Contractor a programme for the execution of the Works for his own consent; to prescribe the form and detail of the programme to the Contractor before submission; to receive from the Contractor a general description of arrangement and method to be used.
14.2. Revised Programme	To request a revised programme if the progress of the Works does not conform to the one accepted.
14.3. Cash Flow Estimate to be Submitted	To receive a detailed cash flow estimate from the Contractor at the beginning of the Works and request a revised forecast as necessary.
15.1. Contractor's Superintendence	To decide the length of time superintendence is required; to approve the person superintending the Works on Site on behalf of the Contractor. The Engineer has the power to withdraw approval of the Contractor's Superintendent and approve his successor.
16.2. Engineer at Liberty to Object	The Engineer can object to the presence on Site of any of the Contractor's personnel who are incompetent, undesirable, negligent or guilty of misconduct. He can consent to the return to Site of any personnel so removed.
17.1. Setting-out	To give the Contractor the basic points of reference for setting-out; to require rectification of errors by the Contractor; in case of any error based on incorrect data supplied by the Engineer, to determine an addition to the Contract Price in accordance with Clause 52 and notify the Contractor and the Employer.
18.1. Boreholes and Exploratory Excavation	The Engineer can require the Contractor to make boreholes or carry out exploratory excavations under Clause 51.
19.1. Safety, Security and Protection of the Environment	The Engineer can require the Contractor to provide and maintain appropriate lights, guards, fencing, warning signs and watching for the protection of the Works and the public.
20.2. Responsibility to Rectify Loss or Damage	To be satisfied with the Contractor's rectification of loss or damage to Permanent Works.

20.3. Loss or Damage Due to Employer's Risks	To require the Contractor to rectify loss or damage to the Works caused by the Employer's Risks (Sub-Clause 20.4), and to determine an addition to the Contract Price in accordance with Clause 52 and notify the Contractor and the Employer.
25.1. Evidence and Terms of Insurances	To be notified by the Contractor when he provides evidence and insurance policies to the Employer.
27.1. Fossils	To be acquainted by the Contractor of the discovery of fossils etc., and to give instructions for dealing with them; to consult the Employer and the Contractor to determine any extension of time or addition to the Contract Price and to notify the Contractor and Employer accordingly.
30.3. Transport of Materials or Plant	To be notified by the Contractor of the occurrence of damage to any roads and bridges; to have an opinion on the responsibility of the Contractor causing damage and to consult with the Employer and Contractor to determine the amount of monies due to the Employer; to notify the Contractor and Employer accordingly.
30.4. Waterborne Traffic	As above if traffic is waterborne.
31.1. Opportunities for Other Contractors	To require the Contractor to give other Contractors, workmen of the Employer and of authorities all reasonable opportunities for carrying out their work.
31.2. Facilities for Other Contractors	To request the Contractor to make available roads, Temporary Works, the Contractor's Equipment or other services to other Contractors and to determine an addition to the Contract Price in accordance with Clause 52 and to notify the Contractor and Employer accordingly.
33.1. Clearance of Site on Completion	To be satisfied with the clearance of Site by the Contractor following issue of Taking-Over Certificates for parts or all of the Site.
35.1. Returns of Labour and Contractor's Equipment	The Engineer may require the Contractor to deliver to him returns in detail in such form and at such frequencies as prescribed by the Engineer showing staff and labour employed on Site and other information about Equipment.
36.1. Quality of Materials, Plant and Workmanship	To give instructions in respect of testing of materials, Plant and workmanship.
36.4. Cost of Tests not Provided for	To be satisfied or otherwise with results of tests required by the Engineer and decide when costs of tests are to be borne by the Contractor.
36.5. Engineer's Determination where Tests not Provided for	To consult the Employer and Contractor to determine the extension of time and costs which is to be added to the Contract Price in respect of tests not covered by Sub-Clause 36.4, and to notify the Contractor and Employer accordingly.
37.1. Inspection of Operations	To have access to the Site, workshops and places where materials or Plant are being manufactured; to authorise other persons to have the same access.

37.2. Inspection and Testing	To be entitled to inspect and test the materials and Plant to be supplied under the Contract.
37.3. Dates for Inspection and Testing	To agree with the Contractor times and places for inspection and testing.
37.4. Rejection	The Engineer may reject materials or Plant if they are defective or not in accordance with the Contract; to notify the Contractor if he does, giving objections and reasons. The Engineer may request re-testing of rejected materials or Plant after the Contractor has made good; to consult the Employer and Contractor to determine costs incurred by the Employer recoverable from the Contractor and to notify the Contractor and Employer accordingly.
37.5. Independent Inspection	The Engineer may delegate inspection and testing to an independent inspector. Not less than 14 days notice of such delegation is to be given to the Contractor.
38.1. Examination of Work before Covering up	To inspect and approve work before it is covered up.
38.2. Uncovering and Making Openings	The Engineer can instruct the Contractor to uncover any part of the Works or make openings and reinstate and make good unsatisfactory work. If work is satisfactory on exposure he has to consult the Employer and Contractor to determine the amount of costs to be added to the Contract Price and notify the Contractor and Employer accordingly.
39.1. Removal of Improper Work, Materials or Plant	The Engineer has the authority to issue instructions for the removal of improper work, materials or Plant.
39.2. Default of Contractor in Compliance	To consult the Employer and Contractor to determine costs recoverable from the Contractor by the Employer if the Employer employs others to remove or substitute or re-execute improper work etc., and to notify the Contractor and Employer.
40.1. Suspension of Work	To instruct the Contractor to suspend progress of Works in such manner considered necessary and to protect and secure the Works.
40.2. Engineer's Determination following suspension	To consult the Employer and Contractor to determine extension of time and cost to be added to the Contract Price incurred by the Contractor in suspending the Works beyond the control of the Contractor and to notify the Contractor and Employer accordingly.
40.3. Suspension lasting more than 84 Days	If the Engineer fails to give permission to the Contractor to resume work within 84 days from date of suspension he can expect notice from the Contractor to require permission within 28 days to proceed with the Works. If permission is not given within this 28 days, the Engineer can expect the Contractor to give notice to treat a part of the Works as an omission under Clause 51, or if the whole of the Works then to expect the Contractor to terminate his employment under Sub-Clause 69.1.

41.1. Commencement of Works	To give notice to the Contractor to commence the Works within the time stated in the Appendix after the date of the Letter of Acceptance.
42.1. Possession of Site and Access Thereto	To receive notice from the Contractor of reasonable proposals for possession of the Site if programme under Clause 14 is not submitted or if there are no requirements in the Contract Documents.
42.2. Failure to Give Possession	To consult the Employer and Contractor to determine extension of time and costs to be added to the Contract Price if the Employer fails to give possession of the Site or access thereto and to notify Contractor and Employer accordingly.
44.1. Extension of Time for Completion	To consult the Employer and Contractor to determine extension of the time due to the Contractor in the event of the circumstances set out occurring and to notify the Contractor and Employer accordingly.
44.2. Contractor to Provide Notification and Detailed Particulars	The Engineer is not bound to make any determination unless the Contractor notifies the Engineer within 28 days after the event arising or within 28 days or other time agreed with the Contractor after such notification of receipt of detailed particulars of any extension the Contractor feels he is entitled to receive.
44.3. Interim Determination of Extension	To take account of the event having a continuing effect on extension of time and to follow the procedure laid down in this Sub-Clause; no decrease in any extensions of time already determined by the Engineer shall result.
45.1. Restriction on Working Hours	To give consent for Works to be carried on at night or on locally recognised days of rest.
46.1. Rate of Progress	To notify the Contractor when the rate of progress is, in the Engineer's opinion, too slow to complete on time; to give consent to the Contractor's proposals to speed up and to working at night and on days of rest; to consult the Employer and Contractor to determine additional supervisory costs incurred by the Employer and to be recovered from the Contractor, and to notify the Contractor and Employer accordingly.
48.1. Taking-Over Certificate	To follow procedure for the issue of a Taking-Over Certificate.
48.2. Taking-Over of Sections or Parts	To follow similar procedure to that in Sub-Clause 48.1 for Taking-Over Certificates for Sections or Parts of the Works.
48.3. Substantial Completion of Parts	The Engineer may issue a Taking-Over Certificate for any part of the Permanent Works.
49.2. Completion of Outstanding Work and Remedying Defects	To instruct the Contractor to amend, reconstruct and remedy defects during the Defects Liability Period or within 14 days of its expiry following an inspection.
49.3. Cost of Remedying Defects	The Engineer can determine an addition to the Contract Price where the Contractor remedies defects etc. not

within the Contractor's control; to notify the Contractor and Employer accordingly.

49.4. Contractor's Failure to Carry Out Instructions

To consult the Employer and Contractor to determine costs incurred by the Employer remedying defects etc. where the Contractor has failed to carry out instructions and to notify the Contractor and Employer of costs to be recovered from the Contractor.

50.1. Contractor to Search

To consult the Employer and Contractor to determine costs to be added to the Contract Price for costs incurred by the Contractor in searching for work which is found to be satisfactory or where defect was not the responsibility of the Contractor, and to notify the Contractor and Employer accordingly.

51.1. Variations

The Engineer can make Variations to the Works as set out in this Sub-Clause. He cannot omit work to be carried out by the Employer or another Contractor.

52.1. Valuation of Variations

The Engineer must investigate whether rates and prices set out in the Contract are applicable; to consult the Employer and Contractor to agree suitable rates and prices to value varied work where existing rates are not applicable. In the event of disagreement he will fix such other rates or prices which are appropriate and notify the Contractor and Employer accordingly; to determine provisional rates and prices to enable interim certificates to be paid until new rates and prices are agreed or fixed.

52.2. Power of Engineer to Fix Rates

To consult the Employer and Contractor to agree suitable rates and prices when any original rates and prices are rendered inappropriate or inapplicable because of Variations; in the event of disagreement to fix other rates and prices as appropriate and to notify the Contractor and Employer accordingly; to determine provisional rates and prices for interim payments. Within 14 days of instruction to vary the Works, no varied work will be valued under Sub-Clause 52.1 or this Sub-Clause unless

(*a*) notice has been received from the Contractor of his intention to claim extra payment or a varied rate or price

(*b*) notice has been given to the Contractor that it is intended to vary a rate or price.

52.3. Variations Exceeding 15 per cent

To consult the Employer and Contractor to determine further sums to be added to or deducted from the Contract Price when variations, as defined, exceed 15 per cent (or other figure agreed in Part II) and to notify the Contractor and Employer accordingly.

52.4. Daywork

To follow procedures laid down to execute work on a daywork basis.

53.1. Notice of Claims

To receive notice from the Contractor within 28 days of an event giving rise to a claim.

53.2. Contemporary Records	To inspect contemporary records on receipt of notice; to instruct the Contractor to keep additional records and call for copies of such records.
53.3. Substantiation of Claims	To follow procedure laid down in this Sub-Clause.
53.4. Failure to Comply	The Engineer is empowered to assess the amount the Contractor is entitled to on the basis of records submitted even if the Contractor fails to comply with the requirements of Clause 53.
53.5. Payment of Claims	To consult the Employer and Contractor to determine the amount due to the Contractor in any interim payment; to notify the Contractor and Employer accordingly.
54.1. Contractor's Equipment, Temporary Works and Materials; Exclusive Use for the Works	To give consent to the Contractor to remove Equipment etc. from the Site.
54.8. Approval of Materials not Implied	The Engineer can reject materials etc. at any time.
56.1. Works to be Measured	To follow procedure laid down in this Sub-Clause.
57.2. Breakdown of Lump Sum Items	To receive and approve a breakdown of lump sums within 28 days of receipt by the Contractor of the Letter of Acceptance.
58.1. Definition of 'Provisional Sum'	To instruct the Contractor to use all or part of Provisional Sums; to notify the Contractor of any determination under this Sub-Clause with a copy to the Employer.
58.2. Use of Provisional Sums	The Engineer has authority to issue instructions in respect of work covered by these sums.
58.3. Production of Vouchers	To receive evidence of expenditure in respect of these sums.
59.1. Definition of 'Nominated Sub-Contractors'	To nominate, select or approve Sub-Contractors for supply and execution of Works where the Contractor is required to sub-let such work to them.
59.4. Payments to Nominated Sub-Contractors	To instruct the Contractor to pay Nominated Sub-Contractors.
59.5. Certification of Payments to Nominated Sub-Contractors	To follow procedure laid down in this Sub-Clause.
60.1. Monthly Statements	To receive after the end of each month six copies of a statement showing the amounts due to the Contractor up to the end of the month; to prescribe form of statement; to approve the person who will sign the Contractor's statement.
60.2. Monthly Payments	Within 28 days of receipt of the Contractor's Statement to certify to the Employer the amount of payment due to the Contractor subject to provisos stated in this Sub-Clause.
60.3. Payment of Retention Money	To follow procedures laid down in this Sub-Clause.
60.4. Correction of Certificates	The Engineer may correct or modify any interim certificate and omit or reduce the value of the work not executed to his satisfaction.

60.5. Statement at Completion

To receive from the Contractor not later than 84 days after issue of Taking-Over Certificate a statement in the form specified by the Engineer

(a) final value of work done up to the date of the Certificate
(b) further sums the Contractor considers to be due
(c) an estimate of amounts the Contractor considers will become due to him.

The Engineer certifies payment in accordance with Sub-Clause 60.2.

60.6. Final Statement

To receive from the Contractor not later than 56 days after issue of the Defects Liability Certificate a draft Final Statement in form approved by the Engineer

(a) final value of sums due
(b) any further sums.

The Engineer may require and agree further information from the Contractor and changes in the draft; to receive the Final Statement from the Contractor.

60.8. Final Certificate

Within 28 days of receipt of the Final Statement and the written Discharge (Sub-Clause 60.7), to issue to the Employer, with a copy to the Contractor, the Final Certificate.

62.1. Defects Liability Certificate

To sign and deliver to the Employer a Defects Liability Certificate in accordance with the procedure laid down in this Sub-Clause.

63.1. Default of Contractor

The Engineer certifies to the Employer that the Contractor, in his opinion, has done or failed to do various actions specified in this Sub-Clause.

63.2. Valuation at Date of Termination

To fix and determine *ex parte* the values and amounts described in this Sub-Clause.

63.4. Assignment of Benefit of Agreement

To instruct the Contractor to assign to the Employer the benefits of agreements to supply goods or materials or services.

64.1. Urgent Remedial Work

To give to the Employer an opinion of work urgently necessary to be executed by or on behalf of the Employer to remedy defects or complete other works. If in the opinion of the Engineer the Contractor was liable to do this work, the Engineer must consult the Employer and the Contractor to determine the costs recoverable by the Employer from the Contractor and notify the Contractor and Employer accordingly; to give notice to the Contractor of the occurrence of the emergency as soon as is practicable.

65.3. Damage to Works by Special Risks

To determine an addition to the Contract Price in accordance with Clause 52 if the Works, materials or plant on or near in transit sustain damage by reason of any special risks.

65.5. Increased Costs arising from Special Risks	To be notified by the Contractor of any costs arising from the outbreak of war; to consult the Employer and Contractor to determine the amount of the Contractor's costs to be added to the Contract Price and to notify the Contractor and Employer accordingly.
65.8. Payment if Contract Terminated	To consult the Employer and Contractor to determine any sums payable under this Sub-Clause and to notify the Contractor and Employer accordingly.
67.1. Engineer's Decision	To receive in writing a reference for a decision from either the Employer or the Contractor in respect of a dispute; to give notice of this decision to both Parties no later than the eighty-fourth day following receipt of the reference. Such a decision must state that it was made subject to this Clause; to receive a copy from either Party of the intention to commence Arbitration.
67.3. Arbitration	The Engineer is not disqualified from being called as a witness or from giving evidence.
68.1. Notice to Contractor; 68.2. Notice to Employer and Engineer; 68.3. Change of Address	To follow procedure laid down for serving and receiving of notices set out in Clause 68.
69.1. Default of Employer	To receive a copy of the notice from the Contractor to the Employer terminating employment under this Clause.
69.4. Contractor's Entitlement to Suspend Work	To consult the Employer and Contractor to determine extension of time and costs to be added to the Contract Price if the Contractor suspends or delays work because of any failure of the Employer to pay amounts due under any certificate of the Engineer and to notify the Contractor and Employer accordingly.
70.2. Subsequent Legislation	To consult the Employer and Contractor to determine the amount to be added to or deducted from the Contract Price as a result of changes to Statutes, Ordinances, Decrees or Laws 28 days prior to submission of Tenders and to notify the Contractor and Employer accordingly.

Responsibility of Contractor

1.5. Notices, Consents, Approvals, Certificates and Determinations	To ensure all notices etc. are in writing and not unreasonably withheld or delayed.
2.1. Engineer's Duties and Authority	(c) The Engineer has no authority to relieve the Contractor of any of his obligations unless expressly stated in the Contract.
2.3. Engineer's Authority to Delegate	To receive a copy of the Engineer's delegation or revocation of duties and authority to be undertaken by the Engineer's Representative.
	(b) The Contractor can query any communication of the Engineer's Representative and refer the matter to the Engineer.

25

2.4. Appointment of Assistants	To receive from the Engineer the names, duties and scope of authority of the assistants.
2.5. Instructions in Writing	To comply with instructions given orally by the Engineer as well as those given in writing. The Contractor can confirm such instructions in writing within 7 days of receiving oral instruction. If confirmation is not contradicted in writing within 7 days by the Engineer, it is deemed to be an instruction in writing. The same applies to instructions given by the Engineer's Representative and any assistants appointed under Sub-Clause 2.4.
3.1. Assignment of Contract	The Contractor cannot assign the Contract without prior approval of the Employer. He can assign a charge in favour of his bankers of any monies due or to become due and assign to his insurers his right to obtain relief against any other party liable.
4.1. Sub-Contracting	The Contractor cannot sub-contract the whole of the Works; unless otherwise provided he cannot sub-contract any part of the Works without prior consent of the Engineer. He does not need consent for (a) provision of labour (b) purchase of materials (c) sub-contracting of any part of the Works for which a Sub-Contractor is named in the Contract.
4.2. Assignment of Sub-Contractors' Obligations	To assign to the Employer at the Employer's request and cost the benefit of any continuing obligation of a Sub-Contractor beyond the Defects Liability Period.
5.2. Priority of Contract Documents	To receive instructions from the Engineer about explanations and adjustments arising from ambiguities and discrepancies in the Contract Documents.
6.1. Custody and Supply of Drawings and Documents	To provide copies of Drawings in excess of two; to return all Drawings, Specifications and other documents to the Engineer on issue of the Defects Liability Certificate. The Contractor cannot provide Drawings etc. to others without the consent of the Engineer unless it is strictly necessary for the purposes of the Contract; to supply to the Engineer four copies of all Drawings, Specifications and other documents submitted to the Engineer and approved by the Engineer, in accordance with Clause 7.
6.2. One Copy of Drawings to be Kept on Site	One copy of all Drawings provided by the Engineer must be kept by the Contractor on Site for inspection and use by the Engineer or by others authorised by the Engineer in writing.
6.3. Disruption of Progress	To give notice within a reasonable time to the Engineer, with a copy to the Employer, whenever planning or execution of the Works may be delayed or disrupted unless the Engineer issues any further Drawings or instructions. Details of Drawings and instruction required, why and by when, and of delays or disruption arising must be given.

6.4. Delays and Cost of Delay of Drawings

If the Contractor suffers delay or incurs cost as a result of the failure or inability of the Engineer to follow notice given under Sub-Clause 6.3, then the Engineer will, after consultation with the Employer and Contractor, determine extensions in time and costs to be added to the Contract Price; the Contractor will be notified with a copy to the Employer.

6.5. Failure by Contractor to Submit Drawings

To be liable for costs if his failure to submit Drawings etc. causes the Engineer to fail to submit Drawings etc. The Engineer will take such failures into account when making the determination authorised in Sub-Clause 6.4.

7.1. Supplementary Drawings and Instructions

To carry out and be bound by any supplementary Drawings and instructions issued by the Engineer necessary for the execution and completion of the Works and the remedying of defects.

7.2. Permanent Works Designed by Contractor

To submit to the Engineer for approval in respect of Permanent Works designed by him, under the Contract, Drawings, Specifications, calculations and other information necessary for the Engineer to be satisfied; to submit operation and maintenance manuals plus Drawings of Permanent Works as completed to enable the Employer to operate, maintain, dismantle, reassemble and adjust such Permanent Works. Completion will not be approved by the Engineer until such information has been submitted to and approved by the Engineer.

7.3. Responsibility Unaffected by Approval

Approval under Sub-Clause 7.2 does not relieve the Contractor of his responsibilities.

8.1. Contractor's General Responsibilities

With due care and diligence, to design, to the extent required, execute and complete the Works and remedy any defects all in accordance with the Contract; to provide superintendence, labour, materials, Plant, Contractor's equipment and all other things necessary.

8.2. Site Operations and Methods of Construction

To take full responsibility for the adequacy, stability and safety of all Site operations and methods of construction; the Contractor is not responsible for design and specification for Permanent or Temporary Works prepared by others, but where the Contract requires him to design Permanent Works he is fully responsible, notwithstanding approval by the Engineer.

9.1. Contract Agreement

If called upon to do so, to enter into and execute the Contract Agreement prepared and completed at the cost of the Employer in the form annexed to these conditions or modified as may be necessary.

10.1. Performance Security (Part II gives example forms including types and proportions of currencies stated in the Appendix to the Tender)

If required, to obtain security for the proper performance of the Contract and provide it to the Employer within 28 days after receipt of the Letter of Acceptance in the sum stated in the Appendix to the Tender; to notify the Engineer in so doing. Such security will be in a form agreed by the Employer and the Contractor. The institution providing the security will be approved by the Employer and the cost will be borne by the Contractor.

10.2. Period of Validity of Performance Security

This security will be valid until the Contractor has executed and completed the Works and remedied defects; no claim can be made after the issue of the Defects Liability Certificate and the security will be returned to the Contractor within 14 days after the issue of this Certificate.

10.3.. Claims under Performance Security

To expect notification from the Employer stating the nature of the default in respect of which a claim is being made.

10.4. (Part II) Source of Performance Security

Security provided by a bank must be accepted by the Employer.

11.1. Inspection of Site (See Part II for Dredging and Reclamation Work)

To interpret data on hydrological and sub-surface conditions provided by the Employer. The Contractor will be deemed to have inspected and examined the Site, its surroundings and information available and have satisfied himself, as far as practicable considering cost and time, before tendering, as to

 (*a*) the form, nature and sub-surface state
 (*b*) the hydrological and climatic conditions
 (*c*) the extent and nature of the work and materials necessary for the Works
 (*d*) the means of access and accommodation required and all necessary information which may affect the Tender.

The Tender will be deemed to have been based on the data made available and the data arising from his own inspection and examination.

12.1. Sufficiency of Tender

To be satisfied that his Tender is based on his knowledge of the Site, conditions and circumstances and will cover all his obligations.

12.2. Adverse Physical Obstructions or Conditions (See Part II for Dredging and Reclamation Work)

To give notice to the Engineer, with a copy to the Employer, of the encountering of physical obstructions or conditions not foreseeable by an experienced Contractor; to follow procedures laid down and to be consulted with the Employer by the Engineer to determine any extensions of time or any costs to be added to the Contract Price; to receive instructions from the Engineer.

13.1. Work to be in Accordance With Contract

To execute the Works in accordance with the Contract to the satisfaction of the Engineer; to comply with and adhere to the instructions from the Engineer; to take instructions only from the Engineer or from the Engineer's Representative (Clause 2).

14.1. Programme to be Submitted

To submit, within the time stated in Part II, to the Engineer for his consent a programme in such detail and form required by the Engineer for the execution of the Works; to provide in writing a general description of the arrangement and method proposed for the execution of the Works.

14.2. Revised Programme	To produce at the request of the Engineer a revised programme necessary to ensure completion within Time.
14.3. Cash Flow Estimate to be Submitted	Within the time stated in Part II, to provide the Engineer, for his information, with a detailed cash flow estimate in quarterly periods and to subsequently revise it if required by the Engineer.
14.4. Contractor not Relieved of Duties or Responsibilities	By submitting such programmes, general descriptions and cash flows, the Contractor is not relieved of any duties or responsibilities under the Contract.
15.1. Contractor's Superintendence (See Part II Sub-Clause 15.2 Language Ability of Contractor's Representative)	To provide all necessary superintendence for the fulfilment of all his responsibilities under the Contract; to provide a competent and authorised representative approved by the Engineer, who will give his whole time to the Works. If approval is withdrawn another supervisor must be provided.
16.1. Contractor's Employees	To provide competent supervisors, technical assistants, foremen, leading hands, skilled, semi-skilled and unskilled labour necessary for the fulfilment of his obligations under the Contract.
16.2. Engineer at Liberty to Object (See Part II Sub-Clause 16.3 Language Ability of Superintending Staff; 16.4 Employment of Local Personnel)	To remove from Site any persons objected to by the Engineer. No one will be allowed back without the Engineer's consent and persons removed will be replaced.
17.1. Setting-out	To be responsible for the accurate setting-out of the Works in relation to the data provided by the Engineer in writing; any errors will be rectified at the cost of the Contractor unless arising from incorrect data provided by the Engineer; to protect and preserve all bench-marks etc.
18.1. Boreholes and Exploratory Excavation (See Part II for Dredging and Reclamation Work)	To make boreholes and exploratory excavations required by the Engineer under Clause 51 unless otherwise provided for in the Bills of Quantities or by Provisional Sums.
19.1. Safety, Security and Protection of the Environment (See Part II for Dredging and Reclamation Work)	To be responsible for safety throughout the execution of the works, completion and remedying of defects; to provide lights, guards, fencing, warning signs and watching.
20.1. Care of Works	To take full responsibility for the care of the Works, materials and plant from Commencement until the date of issue of the Taking-Over Certificate, subject to the Taking Over of Parts of the Works etc.
20.2. Responsibility to Rectify Loss or Damage	To be responsible for loss or damage to the Works as laid down in this Sub-Clause.
20.3. Loss or Damage Due to Employer's Risks	To rectify loss or damage as required by the Engineer in accordance with Clause 52; to accept proportional responsibility if it applies.

21.1. Insurance of Works and Contractor's Equipment	To insure Works as required under this Clause. (See Part II about payment in foreign currencies.)
21.2. Scope of Cover (See Part II for Clauses 21, 23 and 25)	Defines cover to be provided by the Contractor in the names of the Employer and the Contractor.
21.3. Responsibility for Amounts not Recovered	To be prepared to bear amounts not insured.
21.4. Exclusions	To note exclusions.
22.1. Damage to Persons and Property	To indemnify the Employer as provided.
22.2. Exceptions	To note exceptions.
22.3. Indemnity by Employer	To be indemnified by the Employer.
23.1. Third Party Insurance (including Employer's Property)	To insure in the joint names of the Employer and the Contractor against liabilities described in this Sub-Clause.
23.2. Minimum Amount of Insurance	To note the amount stated in the Appendix to the Tender but to consider if this sum is adequate in the circumstances.
23.3. Cross Liabilities	To include a cross-liability Clause as described.
24.1. Accident or Injury to Workmen	To indemnify the Employer against all damages etc. to all persons in his employ or in the employ of Sub-Contractors.
24.2. Insurance Against Accident to Workmen	To take out insurance against liability for accident for injury to workmen as provided.
25.1. Evidence of Terms of Insurances	To provide evidence to the Employer prior to start of work at the Site as required by this Clause.
25.2. Adequacy of Insurances	To ensure the adequacy of insurance provided under Sub-Clause 24.2; to produce to the Employer the policies in force and receipts for payments of premiums.
25.3. Remedy on Contractor's Failure to Insure	To be prepared for the Employer to deduct monies from sums due to cover inadequacies in the Contractor's insurance.
25.4. Compliance with Policy Conditions	To indemnify the Employer against losses and claims arising from failure to insure.
26.1. Compliance with Statutes, Regulations	To comply with Statutes, Laws, regulations etc.; to keep the Employer indemnified against penalites for breach of such Statutes etc.
27.1. Fossils	To follow procedures laid down in this Clause.
28.1. Patent Rights	To indemnify the Employer against any claim for infringement of patent rights etc. as provided by this Sub-Clause.
28.2. Royalties (See Part II for Dredging and Reclamation Work)	To pay all tonnage, royalties, rents etc. for acquiring stone and other materials.
29.1. Interference with Traffic and Adjoining Properties	Not to interfere with the convenience of the public, access to and use of public and private roads etc.; to indemnify the Employer in respect of all claims etc.
30.1. Avoidance of Damage to Roads	To use every reasonable means to avoid damage to roads and bridges by traffic.

30.2. Transport of Contractor's Equipment or Temporary Works	Unless otherwise provided by the Contract, to pay cost of strengthening bridges and improving roads and to be responsible for this work to transport equipment etc. to Site; to indemnify the Employer against all claims for damage to roads and bridges; to negotiate and pay claims made directly to the Employer.
30.3. Transport of Materials or Plant	To notify the Engineer, with a copy to the Employer, as soon as he becomes aware of any damages or receives claims against him. The Employer is not liable for any costs. In some cases the Employer negotiates and pays costs and indemnifies the Contractor. If the Engineer decides that the Contractor has failed to observe his obligations, the Engineer will determine, after consultation with the Employer and Contractor, how much money can be deducted by the Employer from sums due to the Contractor; the Engineer to notify the Contractor accordingly, with a copy to the Employer. However, the Employer must notify the Contractor whenever a settlement is to be negotiated, and if money is due from the Contractor the Employer will consult the Contractor before agreeing any settlement.
30.4. Waterborne Traffic	As above if traffic is waterborne.
31.1. Opportunities for Other Contractors (See Part II Clause 31)	To afford reasonable opportunities to others for carrying out their work as required by the Engineer.
31.2. Facilities for Other Contractors	After being requested to do so in writing by the Engineer, to make available to others various facilities and services; the Engineer will determine additions to the Contract Price under Clause 52 and notify the Contractor, with a copy to the Employer.
32.1. Contractor to Keep Site Clear	To keep the Site clear during the execution of the Works.
33.1. Clearance of Site on Completion	Upon issue of Taking-Over Certificate, to clear away and remove from the Site referred to by the Certificate, all equipment, surplus material, rubbish etc. and leave the Site and Works clean to the satisfaction of the Engineer; the Contractor can retain on Site certain materials, equipment etc. required for work to be done during the Defects Liability Period.
34.1. Engagement of Staff and Labour (See Part II Clause 34)	To make his own arrangements for engaging staff and labour and for their payment, housing, feeding and transport.
35.1. Returns of Labour and Contractor's Equipment (see Part II Clause 35)	To submit appropriate returns to the Engineer giving details of staff, labour and equipment on Site if required by the Engineer to do so.
36.1. Quality of Materials, Plant and Workmanship	All materials, Plant and workmanship must be as specified by the Contractor and the Engineer; to provide resources required for examining, measuring and testing materials and Plant and supply samples of materials for testing before incorporation in the Works.
36.2. Cost of Samples	To bear costs of all samples to be supplied.

36.3. Cost of Tests	To be responsible for the costs of testing required by the Contract.
36.4. Cost of Tests not Provided for	To pay the cost of tests, required by the Engineer, and of the nature set out in this Sub-Clause, if materials, plant or workmanship are not up to standard. (If satisfactory Sub-Clause 36.5 applies.)
36.5. Engineer's Determination where Tests not Provided for	If this Sub-Clause applies, the Contractor and the Employer will be consulted by the Engineer who will determine any extension of time due or additions to the Contract Price caused by the tests not provided for and he will notify the Contractor, with a copy to the Employer.
37.1. Inspection of Operations	To give facilities and assistance to the Engineer to obtain right of access to places where materials and Plant are being manufactured etc.
37.2. Inspection and Testing	To obtain permission for the Engineer to inspect and test materials etc. where manufacture etc. is being carried out on premises not belonging to the Contractor.
37.3. Dates for Inspection and Testing	To agree with the Engineer times and places for inspection and testing. Certain procedures are laid down to be followed by the Contractor and the Engineer.
37.4. Rejection	To be told by the Engineer whenever materials etc. were not ready for testing or failed such tests; to make good any defects and to ensure that any rejected materials or plant are up to the standard required. The Engineer can request tests to be repeated and in this case the Contractor and the Empoyer will be consulted by the Engineer so he can determine how much money is to be recovered from the Contractor. He will notify the Contractor, with a copy to the Employer.
38.1. Examination of Work before Covering Up	To give full opportunity to the Engineer to examine and measure any part of the Works before it is covered, put out of view or have any Works placed upon it; to give appropriate notice to the Engineer to suit.
38.2. Uncovering and Making Openings	To uncover or open up parts of the Works as instructed by the Engineer; the Contractor will pay the costs of opening up if the work is unsatisfactory or faulty. If the work is satisfactory the Contractor and the Employer will be consulted by the Engineer to determine the Contractor's costs involved in uncovering and making good to be added to the Contract Price. The Contractor will be notified, with a copy to the Employer.
39.1. Removal of Improper Work, Materials or Plant	The Contractor can be instructed by the Engineer to remove and rectify improper work, materials or plant.
39.2. Default of Contractor in Compliance	If the Contractor fails to follow the Engineer's instruction authorised in Sub-Clause 34.1, the Employer can arrange for the work to be done by others. The Contractor and the Employer will be consulted by the Engineer to determine the amount recoverable from the Contractor

by the Employer. The Contractor will be notified, with a copy to the Employer.

40.1. Suspension of Work (See Part II Clause 40 when Dredging and Reclamation Work involved)	The Contractor can be instructed by the Engineer to suspend the progress of the Works or of any part, for a time and in a manner considered necessary by the Engineer. During the suspension the Contractor must protect and secure the Works to the degree required by the Engineer. Certain circumstances are described which, if applicable, mean that the Contractor pays the costs of the suspension. (If otherwise Sub-Clause 40.2 applies.)
40.2. Engineer's Determination following Suspension	If the circumstances in Sub-Clause 40.1 do not apply, the Contractor and the Employer will be consulted by the Engineer and he will determine any extensions of time and amounts to be added to the Contract Price. The Contractor will be notified, with a copy to the Employer.
40.3. Suspension lasting more than 84 Days	If the suspension lasts for 84 days the Contractor can give notice to the Engineer requiring possession to proceed within 28 days of the receipt of the notice by the Engineer. If permission is not granted within 28 days, and if only part of the Works is affected, the Contractor can, but is not bound to, treat the suspension as an omission under Clause 51 and give further notice to the Engineer. If it affects the whole of the Works he can treat the suspension as a default by the Employer and terminate his employment under the Contract in accordance with Sub-Clause 69.1, whereupon Sub-Clauses 69.2 and 69.3 will apply.
41.1. Commencement of Works	To commence the Works as soon as practicable after receipt of a notice from the Engineer issued within the time stated in the Appendix to the Tender after the date of the Letter of Acceptance; once started he must get on with the Works expeditiously and without delay.
42.1. Possession of Site and Access Thereto	To be given possession of the Site and access to it by the Employer with the Engineer's notice to commence. Various circumstances are described and in one case the Contractor may have to make proposals to the Engineer, with a copy to the Employer, about his requirements for possession and access. He will be given possession and access to suit the requirements of the Contract and/or his programme for the Works.
42.2. Failure to Give Possession	If he suffers delay or incurs costs because the Employer has failed to give possession and access as required, the Contractor and Employer will be consulted by the Engineer who will determine any extensions of time and costs to be added to the Contract Price and notify the Contractor, with a copy to the Employer.
42.3. Wayleaves and Facilities	To bear all costs for special or temporary wayleaves for access; to provide his own additional facilities required off site.

43.1. Time for Completion (See Part II Clause 43 when Completion may have to be on a certain date)

Times for Completion of the Works or Sections of the Works are stated in the Appendix to the Tender calculated from the Commencement Date and must allow for any extensions allowed under Clause 44.

44.1. Extension of Time for Completion

Certain events are described which may entitle the Contractor to an extension of Time for Completion; in the event he and the Employer will be consulted by the Engineer and the Engineer will determine the amount of the extension and notify the Contractor, with a copy to the Employer.

44.2. Contractor to Provide Notification and Detailed Particulars

To follow the procedures laid down in this Sub-Clause to get the Engineer to determine any extension of Time for Completion.

44.3. Interim Determination of Extension

To follow other procedures laid down where an event has a continuing effect.

45.1. Restriction on Working Hours (See Part II Clause 45 covering isolated areas and Dredging and Reclamation Work)

To seek the consent of the Engineer if he wishes to undertake work at night or on days of rest; emergencies and shift work are covered in this Sub-Clause.

46.1. Rate of Progress

He can be asked by the Engineer to speed up the progress of the Works in certain circumstances and will be unable to claim additional costs; he can seek the Engineer's consent to work at night or on days of rest. If this extra work causes the Employer to spend more money in supervision then these costs will be deducted by the Employer from monies due to the Contractor as determined by the Engineer after consultation with the Employer and the Contractor. The Contractor will be notified, with a copy to the Employer.

47.1. Liquidated Damages for Delay (see Part II Clause 47 for possible bonuses for early completion)

To be prepared to pay liquidated damages as set out in the Appendix to the Tender for the whole, Sections or Parts of the Works. The Employer can deduct such damages from monies due or to become due to the Contractor.

47.2. Reduction of Liquidated Damages

Completion of Sections or Parts of the Works, with appropriate Taking-Over Certificates, prior to the time of completion, enable liquidated damages to be reduced *pro rata*.

48.1. Taking-Over Certificate

To give notice to the Engineer, with a copy to the Employer, with an undertaking to complete any outstanding works during the Defects Liability Period when the whole of the Works have been substantially completed. The Engineer either issues the Taking-Over Certificate within 21 days or instructs the Contractor to do certain works before he can issue it. Certain defects which appear after this instruction and prior to completion must be rectified before completion, 21 days after which the certificate is issued.

48.2. Taking-Over of Sections or Parts

The Contractor may request the Engineer to issue a Taking-Over Certificate for any Sections or Parts of the

Works in circumstances described in (a), (b) and (c) of this Sub-Clause.

48.5. (Part II) Prevention from Testing	Procedure when it is foreseen that the Contractor may be prevented from carrying out tests.
49.2. Completion of Outstanding Work and Remedying Defects	To complete the work outstanding on the date stated in the Taking-Over Certificate; to execute all work of amendment, reconstruction and remedying defects as specified.
49.3. Cost of Remedying Defects	To be responsible for the cost of work undertaken and described in Sub-Clause 49.2 if covered by (a), (b) and (c) of this Sub-Clause; if the work concerned is due to any other cause the Contractor will be paid an amount determined by the Engineer.
49.4. Contractor's Failure to Carry Out Instructions	In case of failure by the Contractor, the Employer is entitled to employ and pay others to do completion and remedial works. The Engineer, after consultation with the Employer and Contractor, will determine the costs to be paid by the Contractor to the Employer.
50.1. Contractor to Search (see Part II for Dredging and Reclamation Work)	The Contractor may be instructed by the Engineer to search for the cause of defect, shrinkage or other fault. If the defect, shrinkage or other fault is not the fault of the Contractor the Engineer will consult the Employer and the Contractor to determine the amount to be added to the Contract Price. If the Contractor is at fault he will bear the cost of the search and of remedying any defect, shrinkage or other fault.
51.1. Variations (see Part II for Dredging and Reclamation Work)	To follow the instructions of the Engineer to make any Variations as defined. The effect will be valued under Clause 52 unless the Variation has been necessitated by default or breach by the Contractor in which case he will bear the cost of such Variation.
51.2. Instructions for Variations	The Contractor will not make any Variations unless instructed to do so by the Engineer.
52.1. Valuation of Variations	To be consulted, along with the Employer, by the Engineer, to agree suitable rates and prices, when the Contract does not contain rates and prices applicable to the varied work.
52.2. Power of Engineer to Fix Rates	To be consulted, along with the Employer, by the Engineer, to agree suitable rates and prices when existing rates and prices are rendered inappropriate or inapplicable; to give notice to the Engineer, within 14 days of the instruction to vary the Works, of his intention to claim extra payment or a varied rate or price; to receive notice from the Engineer of his intention to vary a rate or price within 14 days of the instruction to vary the Works. In either case no varied work instructed under Clause 51 will be valued unless the notices have been given.

52.3. Variations Exceeding 15 per cent	To be consulted, along with the Employer, by the Engineer, to determine the amount to be added to or deducted from the Contract Price arising from the circumstances described in the Sub-Clause.
52.4. Daywork	The ·Contractor may be instructed by the Engineer to execute work on a daywork basis; to give the Engineer receipts and vouchers as necessary; to deliver daily returns of labour, materials and equipment to the Engineer; at the end of each month to deliver priced statements.
53.1. Notice of Claims	To give notice of intention to claim additional payment to the Engineer within 28 days of the event giving rise to the claim pursuant to any Clause in the Conditions.
53.2. Contemporary Records	To keep contemporary records and follow procedure laid down precisely.
53.3. Substantiation of Claims	To follow the procedures laid down precisely.
53.4. Failure to Comply	To recognise that failure to comply with the provisions of Clause 53 may limit the amount the Engineer or any Arbitrator may award.
53.5. Payment of Claims	To expect to be paid the amounts certified by the Engineer in any interim certificate as determined by the Engineer after due consultation with the Employer and the Contractor.
54.1. Contractor's Equipment, Temporary Works and Materials; Exclusive Use for the Works (see Part II Clause 54)	To ensure that all Equipment etc. will be for the execution of the Works, and that nothing is removed without the consent of the Engineer with a proviso for vehicles transporting staff etc.
54.3. Customs Clearance	To expect the Employer to assist in obtaining customs clearance.
54.4. Re-export of Contractor's Equipment	To expect the Employer to assist in obtaining consent to re-export equipment.
54.5. Conditions of Hire of Contractor's Equipment	To ensure that any hired equipment can be used by the Employer on the same terms if necessary.
54.7. Incorporation of Clause in Sub-Contracts	To ensure that any Sub-Contractor's equipment is treated the same as his own under this Clause.
56.1. Works to be Measured	He or his authorised agent will be notified when any Works are to be measured; to follow procedure laid down in this Sub-Clause.
57.2. Breakdown of Lump Sum Items	To submit to the Engineer within 28 days of the receipt of the Letter of Acceptance a breakdown of each Lump Sum item contained in the Tender to be approved by the Engineer.
58.1. Definition of 'Provisional Sum'	To be notified by the Engineer of any determination under this Sub-Clause.
58.2. Use of Provisional Sums	To be instructed by the Engineer for the execution of work, supply of goods, material or services by himself or by a Nominated Sub-Contractor.

58.3. Production of Vouchers	To produce to the Engineer all quotations, invoices, vouchers, accounts and receipts.
59.1. Definition of 'Nominated Sub-Contractors'	All Nominated Sub-Contractors will be deemed to be Sub-Contractors of the Contractor.
59.2. Nominated Sub-Contractors; Objection to Nomination	The Contractor may raise reasonable objections to Nominated Sub-Contractors as laid down.
59.3. Design Requirements to be Expressly Stated	To be indemnified by Nominated Sub-Contractors in respect of design and specification.
59.4. Payments to Nominated Sub-Contractors	To be entitled to be paid as stated in this Sub-Clause.
59.5. Certification of Payments to Nominated Sub-Contractors	To give proof of payment in respect of previous certificates.
60.1. Monthly Statements	To submit to the Engineer after the end of each month six copies duly signed, of a statement, prescribed by the Engineer, showing the amounts due up to the end of the month in respect of items (a), (b), (c), (d) and (e) described.
60.2. Monthly Payments	To expect the Engineer, within 28 days of receiving such a statement, to certify to the Employer the amount due to the Contractor subject to provisos as stated in the Sub-Clause.
60.3. Payment of Retention Money	To receive one half of the Retention Money in respect of all or Part of the Works as laid down in this Sub-Clause on the issue of Taking-Over Certificates; to receive the second half of the Retention Money upon the issue of the Defects Liability Certificate as laid down.
60.4. Correction of Certificates	To be aware of the Engineer's right to correct or modify interim certificates.
60.5. Statement at Completion	Within 84 days of the issue of the Taking-Over Certificate for the whole of the Works to submit to the Engineer a Statement of Completion as defined in this Sub-Clause.
60.6. Final Statement	Within 56 days of the issue of the Defects Liability Certificate to submit to the Engineer a draft Final Statement as defined in this Sub-Clause.
60.7. Discharge	On submission of this Final Statement, to give to the Employer, with a copy to the Engineer, a written Discharge confirming that the Final Statement represents the full and Final Settlement. This only becomes effective after payment due under the Final Certificate has been made and the performance security returned to the Contractor.
60.8. Final Certificate	The Engineer will issue a Final Certificate within 28 days of receipt of the Final Statement and of the written Discharge.
60.9. Cessation of Employer's Liability	To include all unsettled claims in his Final Statement or lose the right to claim.

60.10. Time for Payment	To expect payment of interim certificates within 28 days of delivery to the Employer of the Engineer's certificate or within 56 days of the delivery of the Final Certificate; to be entitled to interest on any delayed payment at the rate stated in the Appendix to the Tender. The Contractor has an entitlement under Clause 69 if payments are delayed.
(60. See Part II for suggested Clauses to cover payment in different currencies and different countries)	
61.1. Approval only by Defects Liability Certificate	To recognise that approval of the Works is covered only by this Certificate.
62.1. Defects Liability Certificate	To recognise the conditions which must be satisfied before the Engineer will issue his certificate.
62.2. Unfulfilled Obligations	Both Parties remain liable for the fulfilment of any obligation incurred prior to the issue of the Defects Liability Certificate and unperformed at the time of issue.
63.1. Default of Contractor	To expect Termination of the Contract under the circumstances described in this Sub-Clause.
63.2. Valuation at Date of Termination	The Engineer will determine any sums due to the Contractor upon termination.
63.3. Payment after Termination	The Engineer will determine any sums due to the Employer by the Contractor.
63.4. Assignment of Benefit of Agreement	To be instructed by the Engineer, under the procedures described, to assign to the Employer any benefits of any agreements to supply goods, materials or services.
64.1. Urgent Remedial Work	If he is unable or unwilling at any time to do remedial or other work urgently necessary for the safety of the Works as required by the Engineer, he will be liable for the costs of such work or repair done by or on behalf of the Employer. These costs will be determined by the Engineer after consultation with the Employer and the Contractor.
65.1. No Liability for Special Risks	The Contractor is not liable for Special Risks described in Sub-Clause 65.2.
65.3. Damage to Works by Special Risks	To be entitled to payment as defined.
65.5. Increased Costs arising from Special Risks	To notify the Engineer whenever costs arising from Special Risks come to his knowledge. The Employer and the Contractor will be consulted by the Engineer who will then determine additions to the Contract Price.
65.6. Outbreak of War	Unless the Contract is determined, to do his best to complete the Works following an outbreak of war; however, the Employer can terminate the Contract as laid down in this Sub-Clause.
65.7. Removal of Contractor's Equipment on Termination	To remove equipment from the Site on Termination under Sub-Clause 65.6.
65.8. Payment if Contract Terminated	To receive payment in accordance with this Sub-Clause.

66.1. Payment in Event of Release from Performance	To be paid as under Clause 65 if the Contract is frustrated as laid down.
67.1. Engineer's Decision; 67.2. Amicable Settlement; 67.3. Arbitration; 67.4. Failure to Comply with Engineer's Decision	Both Parties to the Contract must follow and respect the precise procedures laid down under Clause 67. (Part II suggests possible variations to the use of ICC Arbitration Rules.)
68.1. Notice to Contractor	To receive all certificates, notices or instructions from the Engineer and the Employer at the places or addresses nominated by him.
68.2. Notice to Employer and Engineer	To send any notices to the addresses nominated in Part II.
68.3. Change of Address	All concerned must notify the others of any changes of address.
69.1. Default of Employer	The Contractor can terminate his employment under the Contract by giving notice to the Employer, with a copy to the Engineer, in the event of the Employer failing to act or acting as laid down under items (a), (b), (c) or (d) in this Sub-Clause.
69.2. Removal of Contractor's Equipment	The Contractor can remove his equipment from the Site 14 days after giving notice under Sub-Clause 69.1.
69.3. Payment on Termination	Payment to the Contractor by the Employer is to be as prescribed under Clause 65 with any loss or damage incurred by the Contractor as a result of the Termination.
69.4. Contractor's Entitlement to Suspend Work	The Contractor may, in the case of non-payment of amounts due under certificates, suspend or reduce the rate of work. A set procedure must be followed. The Engineer consults the Employer and the Contractor to determine extensions in time and additions to the Contract Price if the Contractor incurs costs or suffers delay.
69.5. Resumption of Work	The Contractor can resume normal working if the Employer pays interest on delayed payments, pays the amount due under Sub-Clause 69.4 and if the Contractor has not given notice of Termination.
70.1. Increase or Decrease of Cost	To be aware of possible provisions in Part II.
70.2. Subsequent Legislation (see Part II)	To be consulted, along with the Employer, by the Engineer who will determine additional or reduced costs to be added to or deducted from the Contract Price if the circumstances described in this Sub-Clause occur.
71.1. Currency Restrictions	To be aware of rights for additional payment in circumstances when currency restrictions are changed.
72.2. Currency Proportions (see Part II Clause 72)	To be aware of the conditions laid down in this Sub-Clause in Parts I and II.
72.3. Currencies of Payment for Provisional Sums (see Part II Clause 73)	To be aware of the conditions laid down in this Sub-Clause in Parts I and II.

3. Comparison with the third edition

Changes from third to fourth edition

Present users of the third edition will have little concern about the contents of the new edition other than to examine the various changes and to discover whether these have been made simply to improve the language of the document, or if they are of enough significance to influence their thinking on problems that have arisen in using the third edition.

In general terms, the new edition brings the Employer much more into prominence during the period of construction. This enables him to be more informed on certain matters which were previously the sole responsibility of the Engineer, and more particularly to have some say on matters of finance and payment. There is marked improvement in the clarity and precision of the definitions of a number of important terms used within the Document Clauses. Also, the Contractor is given positive direction as to exactly what he must do if he wishes to pursue matters of claims. It is now his responsibility to follow the stipulated procedures when presenting a claim. Failure to do so, and at the correct time, will mean the claim is rejected no matter how well presented and correct it may be.

The Engineer's duties and responsibilities have changed only slightly. He now has to consult the Employer and the Contractor before making certain decisions. Extra payment and extra time are two of the more important issues. He also has the specific requirement of having to act impartially in his decision-making, which serves to formalise something that has always been part of his conduct.

The exact nature of the changes can be examined Clause by Clause. The numbering of the new edition is fundamentally the same as that of the previous edition, except that claims have a separate identity as Clause 53. The previous Clause 53 is now merged with Clause 54.

The important changes are

1. The Employer no longer has the right to change the Engineer simply by advising the Contractor of this happening. The Engineer appointed carries out the duties listed in the Contract and for its duration. The Engineer named is not necessarily the one who prepared the Drawings and designed the Works. *Sub-Clause 1.1(a)(iv)*

40

2. Consents and approvals, certificates, notices etc. to be given by any person such as the Engineer, Employer, Contractor, Sub-Contractor or Arbitrator are to be in writing and not unreasonably withheld or delayed.

Sub-Clause 1.5

3. Instructions from the Engineer are to be given in writing but where given orally a procedure is now laid down whereby the Contractor can confirm such instructions in writing and if not contradicted in writing by the Engineer are deemed to be his instructions.

Sub-Clause 2.5

4. In the previous edition it could be either the Employer or the Engineer who appointed the Engineer's Representative. It is now the sole duty of the Engineer.

Sub-Clause 2.2

5. 'Assistants' can now be appointed by the Engineer or his Representative. They are without authority to give instructions other than those that are necessary to carry out their duties about which the Contractor shall have been notified.

Sub-Clause 2.4

6. It is now an express requirement of the Engineer to exercise his discretionary authority with impartiality. It will no doubt be argued that for the Engineer to act 'impartially', could be different from his traditional role of acting 'independently'. However, Engineers and Contractors alike recognise that the many diverse and exacting functions performed by the Engineer leave little room for genuine complaint.

Sub-Clause 2.6

7. The several documents forming the Contract are now given a listing of 'Priority' for use when ambiguities or discrepencies arise. Such ambiguities and discrepencies must be explained and adjusted by the Engineer. Depending on the exact nature of his instructions, he will determine whether the Contractor is to receive additional payment and time.

Sub-Clause 5.2

8. The Contractor can now reproduce Drawings and such like only if it is strictly necessary, and no Drawings, Specifications or other documents can be passed to any third party without the prior consent of the Engineer.

Clause 6.1

If the Contractor fails to provide Drawings or other documents which in turn prevents the Engineer himself issuing a Drawing or instruction, this, and any failure by the Contractor will be taken into account when the Engineer assesses the extra cost and any extension of time due to the Contractor.

Sub-Clause 6.5

Where the Contractor is responsible for the design of any Permanent Work, he is required to provide four copies of all Drawings and other documents plus one reproducible copy at his cost. If more copies are required he shall be paid for them.

Sub-Clause 6.1

9. The Contractor will not be given a Taking-Over Certificate until maintenance manuals and Drawings have been provided for the Engineer and approved by him.

Clause 48, Sub-Clause 7.2(b) for Permanent Works designed by the Contractor

10. 'Performance Security' is a different requirement of the Contractor compared to Bonds and Guarantees and is required to be provided by the Contractor within 28 days following the Letter of Acceptance. The exact

Sub-Clause 10.1

nature of the form of security required is to be agreed by the Employer and the Contractor. The cost of such security is borne by the Contractor.

No claims will be made after a Defects Liability Certificate has been issued and the security itself will be returned to the Contractor within 14 days of such a Certificate.

Sub-Clauses 10.1, 10.2 and 10.3

Before making a claim on the Contractor under the Performance Security Guarantee the Employer must in every case explain to the Contractor the nature of the default in question.

11.   Cash Flow Estimate is a new obligation placed on the Contractor who is now required to submit to the Engineer a forecast of his estimated cash flow in quarterly periods and to indicate all payments expected to be due to him.

Sub-Clause 14.3

The first estimate is to be provided within the number of days stated in Part II following the date of the Letter of Acceptance and thereafter a revised estimate is to be provided at quarterly intervals if required by the Engineer.

Sub-Clause 14.3

12.   'Safety, Security and Protection of the Environment' now gives the Contractor a greater responsibility for the safety of all on Site and to conform to all the requirements of all the constituted authorities, and, both on and off the Site, to protect the environment from pollution, noise or other causes arising from his methods of operations.

Sub-Clause 19.1

13.   Employer's Responsibilities is a new obligation for the Employer, relating to his employees for safety and maintaining orderly conditions on Site.

Sub-Clause 19.2

14.   The previous edition dealt with Employer's Risks in one very large paragraph. Although the contents remain much the same, the new edition separates the various risks into individual Sub-Clauses. The most significant change is with the 'forces of nature' risk. This is no longer one which the Contractor 'could not foresee or make reasonable provision for or insure against' but is a risk 'against which an experienced Contractor could not reasonably have been expected to take precautions'.

Sub-Clause 20.4

15.   There is now an additional 15 per cent (or different if so stated in Part II) to cover additional costs in respect of Insurance of Works and Contractor's equipment.

Sub-Clause 21.1(b)

16.   There is now a list of specific insurance exclusions of certain Employer's Risks from coverage of the Contractor's required insurance and an allocation of loss not covered by insurance.

Sub-Clauses 21.3 and 21.4

17.   It is now an obligation of the Contractor to notify the insurers of any changes to the programme of execution of the Works and to ensure the adequacy of insurance cover at all times.

Sub-Clause 25.2

18.   If either the Employer or the Contractor fails to comply with the insurance policy conditions each shall indemnify the other against all losses and claims arising therefrom.

Sub-Clause 25.4

19. With regard to transport of Contractor's Equipment or Temporary Works, changes have been made to the financial responsibilities for strengthening bridges and roads. *Sub-Clause 30.1*

20. There are now much more detailed requirements in Part II for the engagement of staff and labour which can increase the Contractor's obligations and costs. *Clause 34*

21. An additional Sub-Clause now includes 'any delay, impediment or prevention by the Employer' as an acceptable reason for granting an Extension of Time for Completion to the Contractor. *Clause 44.1*

22. There are changes in the procedure for the Contractor to seek Extensions of Time. *Sub-Clause 44.2*

23. Sub-Clause 'Interim Determination of Extension' now considers the effect of a continuing effect, and where it is not practicable for the Contractor to submit the necessary details within the required 28 days. *Sub-Clause 44.3*

24. There is now no Clause 47.3 Bonus for Competition in Part I but a bonus Clause can be found in Part II and dealt with as appropriate. *Part II Sub-Clause 47.3*

25. It is now no longer permissible for the Engineer to make a Variation whereby omitted work is taken from the Contractor to be carried out by the Employer or another Contractor. *Sub-Cause 51.1(b)*

26. There has been a tightening up of the Contractor's obligation to give notice 'of his intention' of seeking more money. The words 'as soon as practical' have been replaced with 'within 14 days of the date of such instruction and ... before the commencement of the varied work'. *Sub-Clause 52.2*

27. A separate new Claims Procedure has been included which is extremely clear as to what is required to be done by the Contractor if he is to succeed with his submission. It also emphasises that if he does not follow the laid-down requirements of this Clause to the letter he will not receive anything, however good a case he might have. *Sub-Clause 53.1, 53.2, 53.3, 53.4 and 53.5*

It also becomes necessary for the Contractor to ensure that all the claims he wishes to make are presented within the time limits set out when considering Statements at Completion and the Final Statement. *Sub-Clauses 60.5 and 60.6*

28. There are new provisions concerning the Contractor's Equipment for hire in the event of determination for reasons of default. *Sub-Clauses 54.5–54.8*

29. Certificates for payment: there is a specific identification given to those items considered necessary for evaluating interim payments which had not previously been listed as such. *Sub-Clause 60.1*

30. There are specific arrangements for the release of Retention Monies and certain restraints on their release to be applied by the Engineer if outstanding work has not been completed. *Sub-Clause 60.3*

31. Time for payment: if the Employer is late making payment to the Contractor he now becomes liable to pay a fixed rate of interest as stated *Sub-Clause 69.1(a)*

in the Appendix to the Tender. This must mean that late payment is no longer a Breach of Contract although it still must remain a default of the Employer. This allows the Contractor to determine his employment with the Employer, to suspend the work or to reduce the rate of progress. *Sub-Clause 60.10*

If the third course of action is adopted, the Contractor is entitled to interest on the late payment and any appropriate extra costs and extensions of time when the Employer subsequently does make payment. *Sub-Clauses 60.10 and 69.1*

32.   As part of the Settlement of Disputes procedure there is a new requirement that there must be an attempt to settle any dispute amicably made by the Parties. How this is done is left to the Parties themselves, but it should not be forgotten that the ICC, to whom Arbitration proceedings are to be referred, also provides a service concerned with settlements by amicable means. *Sub-Clause 67.2*

33.   As adverse physical conditions is among the most used Clauses it is fortunate that it has escaped major changes. It is now in two paragraphs, and it is now the Engineer and not the Engineer's Representative who the Contractor notifies, and the notice given by the Contractor must now be copied to the Employer. *Sub-Clause 12.2*

34.   Possibly the most significant change concerns due consultation. It is now an obligation in a number of Clauses, mostly dealing with extra payment to the Contractor, for the Engineer to have due consultation with the Employer and Contractor before making any determination on a particular subject matter. This gives the Employer more direct knowledge of what is happening on Site rather than relying on reports from the Engineer. It also gives him an official voice in matters concerning any extra payments he must make to the Contractor for any changes and additional work ordered by the Engineer. Some consultations do cover payments from the Contractor to the Employer. In one case only does the Employer consult the Contractor with the Engineer present. *Sub-Clause 30.3*

35.   The disappearance from the front cover of the word 'International', and the foreword in Part I suggests that the Conditions are equally suitable for domestic as well as international Contracts.

Part II: Conditions of particular application

Part I of the FIDIC documents, the General Conditions, is generally intended to apply to any type of Civil Engineering Construction or Dredging Contract without alteration, because the Clauses themselves clearly set out the responsibilities and risks undertaken by the Parties, the manner by which the work is to be performed and the payment made when complete. It does not set out to include all the finer detail of particular Contract requirements. Each and every Contract has certain unique functions, risks, obligations, and such like, so a further set of rules, regulations, procedures and definitions has been created. This is Part II: Conditions of Particular Application with Guidelines for Preparations of

Part II Clauses. It sets out all those matters of information, direction, and procedural importance which apply to a particular Contract.

Part II is not a document prepared by the Contractor but is part of the Contract documents provided for him at the time of Tender. It contains 41 Clauses related to the 77 Clauses in Part I. Reference to the Clauses in Part II can give particular and informative details which thereafter control the application of Part I Clauses or other parts of the tendering documentation.

Part II must be part of the Contract between the Employer and the Contractor as without it certain Clauses in Part I are without proper meaning. It embodies those particular requirements applicable to Dredging Work which, because of its peculiarities related mainly to the high capital costs of dredgers, creates different circumstances and understandings. These need special application for the Contract to be completed successfully. The actual Part II document, when provided as part of the published FIDIC Contract, will not be the same as the one presented to the tendering Contractors either in scope or in its precise nature or detail. Instead, a unique Part II will be prepared by the Employer setting out only those matters of importance to that particular Contract. If there is a need for a more or less selective use of the Clauses in Part I, this will be dealt with.

The following detailed examination of the Clauses selected as guidelines, and which form the published Part II, shows that a number of them are devoted to Dredging and Reclamation Work, whereas others provide the essential data for the Parties and the Engineer to achieve an efficient working relationship between all concerned.

1. Definitions	Part II states the names of the Employer and the Engineer. The named Engineer might not be the same Engineer who prepared the design of the Works or even the Contract documents, but is the one given certain duties and authorities to ensure that the Works are constructed properly and completed on time, etc.
2.1. Engineer's Duties	The Engineer's duties are given in detail in Part 1. He is responsible for a number of duties and the exercising of his skills without the help or guidance of the Employer. However, if the Employer wishes to give specific approval to the carrying out of any of the Engineer's duties before they are exercised, the particular Clauses to which they apply must be stated in Part II.
5.1. Language/s and Law	For most legal Contracts it is necessary for the Parties to have agreed as to which country's Laws are to apply and which languages are to be used. This is most important in those countries which have more than one written or spoken language in common usage as many words can have different meanings in different parts of the country. Where more than one language is used in the Contract documents a Ruling Language must be stated.

5.2. Priority of Contract Documents	An example is provided setting out the listing of the various documents in their order of precedence. In Part I, six categories of documents are listed, but in Part II there are eight categories.
9.1. Contract Agreement	The Agreement in Part I is optional but if it is decided that there should be one and if the one given in Part I is unsuitable, a suitable one should be included here.
10.1. Performance Security	If Performance Security is required, its precise requirements shall be given here with appropriate details of the currencies involved.
10.4. Source of Performance Security	This is an extra Clause where, in the event of the Performance Security being in the form of a bank guarantee, it is to be from a bank located in the country of the Employer, or a foreign bank through a correspondent bank located in the country of the Employer.
11. Inspection of Site	There is an addition to Part I that if it is not practicable for the Employer to give all the data in the Tender Documents and some has to be seen at an office, the Contractor must be told the location of the office (Sub-Clause 11.2)
	For Dredging and Reclamation Work this Clause could be varied in respect of certain geological data.
12.2. Adverse Physical Obstructions or Conditions	Dredging and some types of Reclamation Work may require this Clause to be varied.
14.1. Programme to be Submitted	The number of days within which the programme is to be submitted should be given here.
14.3. Cash Flow Estimate to be Submitted	The number of days within which the cash flow estimate is to be submitted should be stated here.
15.2. Language Ability of Contractor's Representative	If the Contractor's Representative needs to be fluent in a language which is not that of the country in which the Works are being executed, it must be stated here. Otherwise an interpreter is to be made available.
16.3. Language Ability of Superintending Staff	A reasonable proportion of the Contractor's superintending staff shall have a working knowledge of a language named in this Sub-Clause. Otherwise, a sufficient number of competent interpreters should be available to ensure proper transmission of instructions and information.
16.4. Employment of Local Personnel	The Contractor is now encouraged to employ local staff and labour. This is sensible as many people can continue being employed on the Project once the Contractor has left the Site by which time they will be properly trained and understand the nature of the Works.
18.1. Boreholes and Exploratory Excavation	For Dredging and Reclamation Work these may need to be varied.

19.1. Safety, Security, and Protection of the Environment	Special care for Dredging Work should be stated here and for the need to make proper provision for additional wording in the event of pollution becoming a risk.
21.1. Insurance of Works and Contractor's Equipment	If payments are to be made in foreign currency this Clause may be varied to take into account the deductible limits of the Employer's Risks.
21, 23 and 25. Insurances Arranged by Employer	All particular requirements concerning matters of insurance should be stated in Part II.
28.2. Royalties	This Clause may be varied for Dredging and Reclamation Work.
31. Other Contractors	Where the particular requirements of other Contractors on Site are known within reasonable limits at the time of Tender, they must be made known here.
34. Engagement of Staff and Labour	Most of the many special factors concerning the use and employment of local labour which have to apply under the Conditions of the Contract are to be stated here. A number of helpful examples are given should they be needed.
35. Returns of Labour and Contractor's Equipment	If returns to cover particular circumstances are desirable they should be stated here.
40. Suspension	Work concerning Dredging and Reclamation Work may require this Clause, as given in Part I, to be varied. Where suspension exceeds 84 days, other adjustments are necessary but should be referred to as Clause 40.3 and not 40.2.
43.1. Time for Completion	If the time to complete is required to be given as a particular date instead of by a number of days, as given in the Appendix to the Tender, it should be stated in Part II.
45. Working Hours	Where the Contract is in an isolated area, or for Dredging and Reclamation Work, this Clause may be varied.
47.3. Bonus	If a bonus is offered for early Completion of the Works or Sections of the Works, it should be stated in Part II.
48.5. Completion Tests	Where the Contractor is prevented for reasons beyond his control from carrying out the necessary tests on completion, an additional Sub-Clause may be needed.
49.5. Defects Liability Period	Where a Contract includes a high proportion of plant, an additional Sub-Clause may be added concerning a possible extension of the Defects Liability Period. An additional Sub-Clause may be added for Dredging Work.
50.2. Searching	For Dredging Work an additional Sub-Clause may be required when an additional Sub-Clause has been adopted in Sub-Clause 49.5.
51.1. Variations	For Dredging and some kinds of Reclamation Work an additional Sub-Clause may be required.
52.1. Valuation of Variations	Where a provision is made for payment in foreign currency this Clause may need to be varied.

Engineer who will confirm, revise or vary the communication as he sees appropriate.

It is interesting that whereas the Contractor's Employees are to be 'only such technical assistants as are skilled and experienced in their respective callings', it seems the Engineer's Representative can be anyone the Engineer may select and who might possess no suitable qualifications to properly perform the duties delegated to him. *Sub-Clause 16.1*

Where the Engineer's Representative issues an instruction or determination to the Contractor concerning any of the authorities delegated to him, the Contractor is obliged to accept it as if given by the Engineer himself. This raises an interesting situation. If the Engineer later decides that any communication from his Representative is not to be taken into effect or processed, it is not just a simple matter of withdrawing the communication. It could be regarded as a Variation and any such instruction from the Engineer ought to be processed under Clause 51. *Sub-Clause 2.3*

Clause 51

There is, in the running of the Contract, a possibility that the Engineer's Representative will be faced with a workload he is unable to perform alone. The volume or the technical nature of certain aspects of his work requires help from Assistants working for him. Such Assistants may be appointed by the Engineer or his Representative but they can have no authority to issue any instructions to the Contractor other than those necessary for them to carry out the duties delegated to them.

In most major Contracts, the running of the Engineer's office requires a close liaison with the Contractor so that all the contractual and technical functions, by which the Works are to be completed, can be performed efficiently and without unnecessary delay. In particular, the Engineer's Representative must be closely concerned with maintaining the proper flow of money between the Employer and the Contractor and keeping adequate records of daily events so he can play an important part in the settlement of any claims which arise during the construction of the Works.

Engineer to act impartially (Sub-Clause 2.6)

This is a new Sub-Clause in the fourth edition, the obligation of which was only implied in the third edition. In the third edition there were two Clauses in which the Engineer's reasonableness (possibly the same as impartiality) had to be exercised and a few Contractors found it necessary or justifiable to complain that the Engineer did not act with reasonableness throughout the duration of the Contract. However, there have been instances when the Engineer to a Contract did act in a different manner from other Engineers on a similar Project which required the Contractor to adhere to a much more rigid application of the specification than the Contractor himself considered reasonable. *Sub-Clause 2.6*

Some Engineers might feel a little hurt that this new Sub-Clause formally requires them to be seen to act with impartiality, especially when in their opinion they have already done this without question since they started in practice. This is understandable, but the formal basis gives the opportunity for any apparent lack of display of impartiality to become a matter for Arbitration if either Party considers he is not getting impartial treatment under the circumstances which prevailed.

Unfortunately, the Sub-Clause raises questions as to what happens when, according to the Contractor, a decision made by the Engineer is not given impartially. Does the Contractor have to continue with the Works and leave the decision unchallenged or unresolved only to find that over time it has increased his costs considerably? Does he have to act immediately to obtain an Engineer's decision on the injustice he considers is being done to him by the Engineer and quickly seek Arbitration under Clause 67? Or should he appeal directly to the Employer?

Clause 67

An appeal made to the Employer is not as unlikely in the new edition as it would have been under the third edition because certain Clauses now require the Engineer 'to consult with the Employer and the Contractor'. Therefore it would seem appropriate that the Contractor should also claim the right of direct consultation because to accept the Engineer's decision without complaint when it was given could lead to an unnecessary dispute, but by referring it to the Employer such a dispute might be avoided.

It must be accepted that the Engineer is likely to perform his decisions with greater skill and ability than the Employer ever could. But as the Employer is the one to be called upon to pay for the outcome of a decision which is claimed to be partial, then he might want to be involved at an early stage to avoid a costly dispute.

Due consultation

It was a stated intention in the fourth edition of the FIDIC *Conditions of contract* to bring the Employer into more prominence so that he could become more involved in the day to day activities on site and in particular those which involve financing the Contract Works. In the third edition, his involvement with the details of construction was more noticeable by his absence, but under the fourth edition it is intended that he is consulted by the Engineer on a variety of subjects, mostly affecting payment to the Contractor. The fourth edition, however, fails to define exactly what is intended to be the purpose or outcome of the 'consultation' by the Engineer with the Employer and the Contractor, and indeed what status it is to be given. It is neither stated nor implied whether the Engineer is expected to be so influenced by his consultation with the Employer or the Contractor that he changes his mind about what is otherwise intended to be 'determination' by the Engineer. If he is so influenced by either Party can the outcome really be his own determination or decision?

It might be the intention of the fourth edition that, every time the Engineer is obliged to have a consultation, he should ask the Employer and the Contractor to attend at the same time. If this is so, it is not stated. If there is to be a tripartite discussion, with no declared or defined purpose, it is easy to imagine the emotion that would be generated at such an informal meeting if either the Employer or the Contractor were to feel aggrieved if an Engineer does not determine a matter in his favour. If the Engineer were to meet each Party separately before making any determination, the procedure would be no different in the future than it was in the past. Therefore, simply because it is an innovation of the fourth edition, it is likely that he will meet them together. This should certainly apply for significant situations and it would be proper for the Engineer to state that he had seen both Parties when giving his decision.

Most unsatisfactory for the long-suffering Engineer is the possibility of being accused of bias, or that the determination he gives is not genuinely impartial but rather a reflection of the consultation he has had with one Party or the other or both.

There are 23 Clauses which require the Engineer to consult the Employer and the Contractor at the proper time. All but one concerns payment to or from the Contractor, eight can influence the Time for Completion, and part of one contains a requirement for the Employer to consult the Contractor before reaching settlement including a sum due from the Contractor to the road or water authority. Over a number of years, some *Sub-Clause 30.3* Employers involved with large Projects and Contracts, retain for themselves certain rights and involvements in decision-making concerning payment to the Contractor which would otherwise be those of the Engineer. There can be no complaint over such an arrangement as it is usually very clearly *Sub-Clause 2.1* stated in Part II as to what they are. The Contractor is therefore fully aware of the situation at the time of submitting his Tender.

Clauses where the Engineer is obliged to undertake 'due consultation with the Employer and the Contractor' will now be examined.

6.4. Delays (time and cost to Contractor)	These can occur either by the failure of the Engineer to give timely Drawings or instructions, or by the failure of the Employer to provide various materials on time as he undertook to do.
12.2. Physical Conditions (time and cost to Contractor)	The Engineer is testing his opinion that such obstructions or adverse conditions could not have been foreseen by an experienced Contractor.
27.1. Fossils, Antiquities and Structures (time and cost to Contractor)	Assessing impact of Engineer's decision.
30.3. Transport of Materials or Plant (cost to Employer)	Failure of Contractor to perform obligations and isolated instance where the Employer is required to consult the Contractor on a matter of a settlement.
30.4. Waterborne Traffic (cost to Employer)	As for Sub-Clause 30.3. where waterborne transport is used.
36.5. Tests (time and cost to Contractor)	Assessing effects of tests ordered by the Engineer.
37.4. Rejection (cost to Employer)	This concerns the recovery of money from the Contractor by the Employer.
38.2. Uncovering work (cost to Contractor)	Assessing result of uncovering work which is found to be acceptable.
39.2. Default of Contractor (cost to Employer)	This concerns the recovery of money from the Contractor by the Employer.

40.2. Suspension (time and cost to Contractor)	Assessing result of order to suspend work outside the control of the Contractor.
42.2. Failure to Give Possession (time and cost to Contractor)	Caused by failure of the Employer to give possession of all or part of the Site or access thereto.
44.1. Extension of Time (time to Contractor)	Assessing impact of events affecting time outside the control of the Contractor.
46.1. Rate of progress (cost to Employer)	Assessing costs incurred by the Employer because of lack of progress by the Contractor.
49.4. Contractor's failure to carry out instructions (cost to Employer)	Recovery of costs incurred by the Employer if the Contractor is in default.
50.1. Searching (cost to Contractor)	Assessing result of abortive searches when work is satisfactory.
52.1. Fixing new rates (fix new rates)	To establish new rates where none existed for work ordered by the Engineer.
52.2. Adjusting rates (revised rates)	To establish revised rates where Variations ordered by the Engineer affect existing rates.
52.3. Variations exceeding 15 per cent (cost to and from Contractor)	To add or to deduct from the Contract Price.
53.5. Claims (cost to Contractor)	Assessment of claims accepted by the Engineer.
64.1. Urgent remedial work (cost to Employer)	Assessing costs arising from the Contractor's failure to do remedial works.
65.5. Special risks (cost to Contractor)	Assessing the cost to the Contractor arising from the impact of Special Risks.
65.8. Termination due to Special Risks (Net payment to or from Contractor)	Assessing payments due to or from the Contractor.
69.4. Suspension/delay (time and cost to Contractor)	Assessing costs to the Contractor when the Employer fails to pay.
70.2. Subsequent Legislation (cost to and from Contractor)	This is an added or deducted amount according to the outcome of subsequent legislation.

It now appears that the particular manner in which the Employer is to become more active in the application of the Contract is likely to be more positive than at first expected. He is now more aware of the extra expense the Engineer is committing him to and what he ought to say about it at the appropriate time.

The Contractor will no doubt welcome the possibility of closer contact with the Employer, who after all is the one who pays him. He will make the most of this new opportunity to be able to make a direct plea of poverty and deprivation should he fail to receive some extra payment following consultation. Even if he fails here he can at least set out the groundwork to direct the Employer towards an eventual settlement in the form an an *ex-gratia* payment in due course.

All things considered it might be better in the long run for the Engineer to recover his all-powerful status and to make all the necessary engineering judgements and financial settlements without involving the Employer. After all there still remains the right of Arbitration if the Contractor or the

Employer feels that the Engineer has not done his job properly either technically or financially.

Contract Documents (priority, ambiguities and discrepancies)

There are many documents which govern and control the way in which the Contract is to be run. Only those listed in the Tender will apply, although others can be added during the construction of the Works and may become of equal importance, for example Working Drawings, Variations to the Works or Drawings provided by the Contractor. The listed documents are *Sub-Clause 5.2* to be mutually explanatory and are given a listed priority of status. Of course, a 'document' in the context of the Contract itself, as well as the Conditions of Contract, means not only the Bills of Quantities, Specifications and such like, but also Drawings, programmes, permits, *Sub-Clause* licences, schedules of materials and indeed everything upon which the *1.1(b)(iii)* Contract Price is calculated.

Because of the complex nature of the work involved in preparing Tender Documents 'ambiguities or discrepancies' can occur. These need to be resolved quickly and positively by the Engineer. In the event of such ambiguities or discrepancies he has to explain to the Contractor exactly what is intended and to issue appropriate instructions to him as may be necessary. The Contractor has a right to expect that the explanation given by the Engineer accords with the listed priority of the Tender Documents, in view of the fact that it was upon their listing and understanding of priorities that his Tender was based before any ambiguities or discrepancies were detected.

If the explanation given makes it necessary to change the order of such listed priorities, then a change to the Contract could be necessary. Should the Engineer find it necessary to issue any 'instruction' which involves the Contractor in additional work or expense, then such a change or instruction must be dealt with under an appropriate Clause in the Contract, and where appropriate a Variation Order should be given. However, if discrepancies or ambiguities occur within the Contract documents, Part II will prevail over Part I as Part II is about those 'particular applications' which apply to that Contract alone. In Sub-Clause 5.2 of Part II an alternative to giving the documents an order of precedence is proposed. This makes all documents mutually explanatory so there is no order of precedence to be applied.

Any explanations given by the Engineer which relate to the resolution of ambiguities and discrepancies could affect the Contractor in a number of ways, for example

(*a*) variations, additional works and omitted works
(*b*) changes to the Contractor's method of construction
(*c*) delays or suspension
(*d*) matters concerning Sub-Contractors
(*e*) extra expenditure if the Contract period is extended.

Of all the Contract documents, the most important to the Contractor are the Drawings, as without these work cannot proceed according to the programme. Likewise, in their absence, certain operations will just stop.

Drawings are given an extremely wide definition within the Contract, so much so that the understanding in the fourth edition of what constitutes a Drawing is totally different from traditional understanding. It is important, therefore, that the Contractor makes a particular note of the definition of Drawings as given in Sub-Clause 1.1(b)(iii) to ensure that his Tender Price fully covers him for all costs he is likely to incur in this connection.

Sub-Clause 1.1(b)(iii)

Drawings are usually issued by the Engineer. When the Contractor has a different role to play, as when he designs work or is concerned with certain specialist undertakings, he issues Drawings for approval by the Engineer. All Drawings to be issued by the Engineer should be available in sufficient detail, in time, and in such a sequence as to permit the Contractor to proceed with the Works without delay or disruption. If the Engineer fails to do this, the Contractor has to give proper notice to the Engineer as to his requirements. Should it appear to the Contractor that he will incur extra cost or delay if a Drawing or further instruction is not given within a reasonable time by the Engineer, he should give notice to the Engineer, with a copy to the Employer, stating what is required and giving his forecast of any delay or disruption he is likely to suffer. If after all this the Engineer fails, or is unable, to issue any Drawing or instruction within a reasonable time for which the Contractor has given notice, and if the Contractor suffers delay and/or extra cost, the Engineer will, after due consultation with the Employer and the Contractor, give an Extension of Time and determine the amount of such extra costs to be added to the Contract Price. However, if any delay on the part of the Engineer to issue any Drawings or instruction is caused by, in whole or in part, the failure of the Contractor to submit Drawings, Specifications or other documents which are his responsibility under the Contract, the Engineer must take into account such failure when making his determination.

Sub-Clause 6.1

Sub-Clause 6.3

Sub-Clause 6.4

Sub-Clause 6.5

Site and access

Before any construction project is designed it is usual for the Site to be selected and arrangements made for it to be available for work to be performed on it as and when required. This is the responsibility of the Employer, who must not only ensure that this is done, but also that he has some form of legal ownership as well as being satisfied that proper and adequate access is available. Once this is done it is possible for the Site to be subjected to a detailed survey and for a report on the geological nature of the hydrological and sub-surface conditions to be prepared. This survey and report are especially important to the Engineer and the Contractor. The Engineer needs it for design purposes and the Contractor for determining his method of construction and to formulate his Tender Price. All data, obtained either on behalf of or by the Employer, has to

Sub-Clause 11.1

be made available to the Contractor. It is on this data that he is deemed to have based his Tender.

Access to the Site must be adequate but sometimes special or temporary wayleaves need to be obtained. If these are necessary the Contractor bears all costs involved. Also, if the Contractor needs additional land or other facilities outside the defined Site area, he has to negotiate and make the necessary arrangements and remember that neither the wayleaves nor the additional land he obtains are designated parts of the Site.

Sub-Clause 42.3

So what then is the Site? Fortunately there is a definition in the Conditions of Contract which calls it 'the places provided by the Employer where the Works are to be executed and any other places as may be specifically designated in the Contract as forming part of the Site'.

Sub-Clause 1.1(f)(viii)

The view has been expressed that if there was a Contract Drawing No. 1, giving full details of the Site and its access, it would be a much clearer means for the exact identity of the Site to be appreciated by all concerned. It is pleasing to find that some Engineers do exactly this. If the location and the extent of the Site are known, the Contractor has the option of visiting the Site before the Tender, and should he decide not to, then according to the Contract, he is deemed to have done so, and if in the future a problem or claim situation arises he has no grounds for pleading ignorance of the prevailing circumstances.

The prudent Contractor who visits the Site will be able to examine the terrain, the local and surrounding conditions, and the facilities available to him and his staff, then with the data provided by the Employer be able to make his Tender realistic and cover all his obligations should he obtain the Contract.

Sub-Clause 12.1

The Engineer, in accordance with Sub-Clause 41.1, gives the successful Contractor notice to commence the Works within the number of days stated in the Appendix to the Contract, after the date of the Letter of Acceptance — the formal acceptance by the Employer of the Tender. The Contractor must commence the Works as soon as is reasonably possible after receipt of the Engineer's Notice.

Sub-Clause 41.1

For the Contractor to commence operations he must have possession of the Site, or parts of it, and access as set out in Sub-Clause 42. In some cases, the sequence of possession of parts of the Site and access connected to the order of execution of the Works is presented in the Tender Documents provided for the Tenderers. Here the Employer will give the Contractor possession of the Site etc, to suit the predetermined sequence simultaneously with the Engineer's Notice to Commence. If no possession details have been specified in the Tender, then, with the Notice to Commence, the Employer must give the Contractor sufficient area of the Site and access to it to enable him to commence, and to proceed to suit his programme submitted under Clause 14. The Employer must otherwise give immediate possession of parts of the Site and access in accordance with reasonable proposals made by the Contractor to the Engineer and to the Employer under Sub-Clause 42.1. Thereafter possession must be given to the Contractor to suit either his programme or the proposals referred to above.

Sub-Clause 42.1

Clause 14

Sub-Clause 42.1

Programmes

The manner in which the Contractor prepares his estimate for a Tender for a major Civil Engineering Contract is generally by making an overall assumption of estimated time elements and the cost of the use of his resources. This way he can establish the total cost and profit of the various sections of the work involved and then relate them to the quantities contained in the Bills of Quantities. His Tender is prepared generally on the overall concept of involvement of resources and not so much on a build-up of separate and single items of work. This method of estimating becomes very meaningful when it is translated into the form of a construction programme which, to the Contractor, will indicate the respective amounts of earnings he can expect in relation to expenditure and the sequences according to which the work will be performed. This outline programme will enable him to fulfil his obligations for the construction of the Works and to complete the Works within the Time for Completion required by the Tender Documents.

The purpose of such a construction programme is to set out the Contractor's timing of the various stages of the Works in calendar form and also to identify the timing of those matters over which he has no direct control, such as the availability of the Site and access thereto, the production of further Drawings and details by the Engineer, the delivery to the Site of Plant and materials provided by the Employer, and all other important factors which may affect his operations, progress and performance.

The programme, as stated in the Tender Documents, is limited to the number of days forming the Time for Completion, as given in the Appendix. Where the Works are to be completed in Sections or Parts, the number of days for the Time for Completion of such Sections or Parts will also be given. In some cases a Date for Completion is given instead of a number *Part II Clause 43* of days. This is identified in Part II Clause 43, and the Dates for Completion of Sections or Parts are to be stated separately.

The need for a detailed programme to be part of the Tender is not covered by the standard form. The need for the Contractor to submit a programme after a number of days from the date of the Letter of Acceptance is covered by Sub-Clause 14.1. This programme is submitted for consent of the *Sub-Clause 14.1* Engineer, although the giving of his consent is not required within any specified time after submission. The Conditions of Contract do not cover what happens if the Engineer does not give his consent.

It is difficult for a Contractor to anticipate exactly the form and detail that the Engineer might require for such a programme as they can differ from Engineer to Engineer. Some might be satisfied with a traditional bar chart programme whereas others might want to use the most sophisticated computer-controlled systems available. The Contractor has the right, in this matter of programme, to expect the Engineer's requirements to be kept to a normal cost expenditure as might be found in the Specification or elsewhere in the Contract Documents so that any cost over and above these requirements will qualify as an increase to the Contract Price.

What happens when the Contractor has submitted a programme under Sub-Clause 14 to the Engineer based on the form and detail he requires only to find that the Engineer does not consent? The situation will obviously be different if the work has already commenced, but if the Contractor cannot proceed as he is obliged to do, with due expedition and without delay, he must consider his options under other conditions of the Contract. If the programme requires the Engineer's consent for the proper execution of the Works, can the Contractor seek an Engineer's order to suspend the Works while the matter is being resolved? Should he give notice to the Engineer that the works are likely to be delayed or disrupted? Does he have a debate with the Engineer as to whether 'consent' is different from 'approval'? They are referred to separately in Sub-Clause 2.6 which *Sub-Clause 2.6* requires the Engineer to act impartially on both, so it could be argued that they cannot mean the same thing. Possibly the second option is the best but the other can apply depending on the actual circumstances.

Care is taken within the Conditions of Contract to ensure that any consent or approval given by the Engineer does not relieve the Contractor of any of his responsibilities. It is sometimes overlooked, however, that while this is so, the Engineer himself can become involved simply by having given his consent where he also had the opportunity to refuse it.

There is no precise definition as to what constitutes a programme and difficulties can therefore arise. It would seem, however, that the programme need demonstrate only the timing and sequences of the starting and stopping of the important work operations to show how the Contractor intends to complete the Works, and how the sequences are related to the rates contained in the Bills of Quantities.

Once the programme has received the consent of the Engineer it is given an almost sacrosanct status and becomes the yardstick of the Contractor's required performance against which his actual progress is recorded. If this is unsatisfactory, the Engineer can request a further programme designed to enable the Works to be completed on time, but in this case the Contractor is not obliged to obtain the Engineer's consent nor the Engineer to give it.

There is no specific direction given as to what happens if the Engineer does not like the new programme. Presumably the Contractor would have to go on submitting further programmes until the Engineer was satisfied, but the work must neither stop nor slow down while this is happening.

In addition to a programme, the Engineer has the option to ask the *Sub-Clause 14.1* Contractor to give a general description of the arrangement and methods he intends to adopt for the execution of the Works. If this option is exercised, it is directed towards indicating the physical method of performance rather than the timing within which each operation is expected to be performed. It is interesting that in providing this information to the Engineer, his consent or approval does not have to be obtained, but that no further action is required from the Contractor or the Engineer once it has been provided. Should the Engineer not like the arrangements and methods the Contractor intends to adopt to do the work, and upon which he has based his Tender, then inevitably a claims situation will arise if the Contractor is obliged to make changes to accommodate the Engineer's wishes.

It would be naive to believe that every Contractor's programme is truly related to the concept of his estimating processes. After all, it is only when he has entered into a legally binding Contract with the Employer that he is called on to produce such a programme for the Engineer's consent under Sub-Clause 14.1. Therefore it must be considered a possibility that some Contractors might use this opportunity to provide a late disclosure of important data and indicate a much earlier sequential completion of construction operations than the one his Tender was truly based on. A requirement for the Contractor to submit a programme as part of the Tender would prevent this. This would give it immediate Contract Document status and would avoid a lot of the present problems. After all, there is now a legal precedence under English law for claiming that a programme submitted and accepted as part of the Contractor's offer is different from a programme which only comes into being once a legal Contract has been established.

Such a programme accepted as part of the Tender could be used by the Contractor to support a claim that the rates and prices in the Bills of Quantities can become inappropriate or inapplicable if the timing of operations in the programme are affected by actions of the Engineer or the Employer. This assumes that the relationships between the programme and the rates are known and have been accepted. If the Contractor fails to complete the Works within an accepted programmed time which is less than the time stated in the Appendix to the Tender or any extended time, (but nevertheless within the Appendix Time), no liquidated damages can be applied as it is the Appendix Time, with any extensions, for Completion, which applies and not the timing as shown on the programme. The simple difference between a Tender programme and one provided under Clause 14 is that the former has already been accepted by the Employer as part of the Contract, and the latter has only to be subject to the consent of the Engineer, and this being given only after a legal Contract is in existence.

Commencement to taking over

Most legal documents concerned with construction contracts set out to give a clear and precise date as to when work is to commence, how long it is to take and how commencement and completion are to be defined. This becomes especially important in legal definition, because without a contractual beginning there can be no contractual end.

In the FIDIC *Conditions of contract* this is done by considering the time for the Commencement of the Works as from when a notice by the Engineer to the Contractor is given within the number of days from the date of the Letter of Acceptance as specified in the Appendix to the Tender. However, it is the date of receipt of such notice from the Engineer which becomes the official starting date for work to commence, and not the date when it was written. Once it is received by the Contractor he is obliged to proceed

Sub-Clause 41.1

with the Works as soon as is reasonably possible. There is also a prerequisite that the Contractor is to be given by the Employer, and at the same time as the notice from the Engineer, possession and access to as much of the Site as the programme referred to in the Contract might indicate or, in the absence of such a programme, in accordance with such reasonable proposals as the Contractor may submit. If the Contractor does not receive the Notice to Commence from the Engineer within the number of days given in the Appendix to the Tender, he should examine his position and seek to recover any extra costs he incurs and an appropriate Extension of Time for Completion under Sub-Clause 42.2. Although this Clause is not specifically designed for action in this situation, it is probably the best Clause available.

Sub-Clause 14.1

Sub-Clause 42.2

If the Employer fails to make available those parts of the Site necessary for the Contractor to work properly, it is a different matter and is dealt with by the Engineer who, after due consultation with the Employer and the Contractor, will award extra costs and time as considered appropriate to the Contractor.

Sub-Clause 42.2

While it is clear that the Engineer is required to issue his notice to the Contractor within a limited number of days from the Employer's Letter of Acceptance, no reference is made to any time limit within which the Employer is required to send his Letter of Acceptance. Obviously the Contractor's Tender cannot remain open indefinitely, so under Item 4 of his Tender, he agrees to a stated number of days after which the offer lapses unless otherwise agreed. However, the Employer can use all the days stated in Item 4 within which to accept the Tender.

After establishing the official Commencement Date, the Contract makes provision in the Appendix to the Tender for a Time for Completion within which the Contractor undertakes to hand over the Works in the manner required. This is given as a stated number of days although it could equally be given as a specific date if this is the wish of the Employer.

Part II Clause 43.1

Under certain circumstances, the number of days or the specific Date for Completion can be extended by the Engineer. This is acceptable for a number of reasons, such as Variations, delays, or Suspension of the Works, but such an Extension of Time for Completion does not, under this Sub-Clause, entitle the Contractor to additional payment itself. He must rely on the particular circumstances for which an Extension of Time for Completion was given. The number of days given for the Completion of the Works can be increased or left unchanged, but never reduced in the event of work being omitted.

Sub-Clause 44.1

The number of days stated in the Appendix to the Tender are calendar days, not actual working days, and so the Contractor must make due allowance in the programming of his operations to exclude night work and locally recognised days of rest. If the Contractor's rate of progress, according to the programme, falls behind, the Engineer can require the Contractor to take measures to catch up with the programme and under these particular circumstances the Contractor can seek the Engineer's consent to work nights and on days of rest.

Sub-Clause 45.1

Sub-Clause 46.1

It is generally understood that when an offer is made by a Contractor it starts the early part of its legal existence from the moment it is received by the Employer it is addressed to. The formation of a legal Contract is established when a proper and unqualified Letter of Acceptance is posted and not from the date it is received.

The Contractor must not be drawn into believing that a 'Letter of Intent' sent in place of a Letter of Acceptance serves the same purpose. A Letter of Intent is merely an indication of an intention by the Employer to enter into a Contract which might or might not materialise. Of course such a letter can become more helpful and improve in status if it states that the Employer will compensate the Contractor for any expenditure should he not secure the Contract.

The FIDIC *Tendering procedure*[2] is helpful in setting out in detail its opinion of a Letter of Intent and acknowledges that under certain circumstances it might serve a useful purpose. It gives the advice that a Letter of Intent issued as a prelude to entering a formal Contract should contain

(*a*) a statement that it is intended to accept the Tender

(*b*) instructions to proceed (or not to proceed) with certain work e.g. mobilisation, ordering of materials, and letting of Sub-Contracts

(*c*) the basis of payment for work authorised and a limit to the financial liability which may be incurred before formal acceptance of the Tender

(*d*) a statement that if the Tender is subsequently not accepted or the Letter of Intent withdrawn, the costs legitimately incurred by the Contractor will be paid by the Employer

(*e*) a statement that when the Tender is formally accepted with a Letter of Acceptance, the provisions of the Letter of Intent will become void

(*f*) a request to the Contractor to acknowledge receipt of the Letter of Intent and to confirm his acceptance of its conditions.

The completion of the construction of the Works, or Sections, if applicable, including any prescribed tests, is recognised by the Engineer issuing the Contractor with a Taking-Over Certificate. This Certificate accepts the fact that not only is the work completed to his satisfaction but also establishes the date for the transfer of certain insurance and care of the Works responsibilities, the release of part of the Retention Money and the start of the Defects Liabililty Period. Such a certificate makes possible the commencement of the Contractor's evacuation from the Site of his plant, personnel and such like, and the closing down of various facilities which he has been using but leaving behind whatever is necessary to undertake any work reasonably expected to arise during the Defects Liability Period. *Clause 48*

Sub-Clauses 20.1 and 21.2

Sub-Clause 60.3(a)

Sub-Clause 49.1

The Contract is deemed to be effectively complete when a Defects Liability Certificate is issued by the Engineer although there are certain continuing obligations of both Parties which can be extended either because *Sub-Clause 62.1* of the Contract Conditions or simply because of the particular laws and ordinances of the country in which the Contract is being performed.

Sub-Clause 62.2

Variations, alterations, additions and omissions

For anything to be subject to a Variation there must be a basis from which a Variation can arise, and, in the context of civil engineering construction, it is generally referred to in the document as being the 'Scope of the Works'. This describes the nature of the Works, whether it is the construction of a highway, airfield, dam or any other utility, and would immediately identify the type of work to be undertaken and generally give a sufficiently detailed description for the Contractor to understand its boundaries.

The Scope of the Works, once defined, may require numerous alterations to successfully adapt to circumstances which could not necessarily have been foreseen when the Works were first envisaged. Such changes might be requested by the Contractor to suit his construction methods, or by the Engineer when producing his Working Drawings. Sometimes the Employer initiates changes to suit his financial or commercial circumstances or to take advantage of technical improvements or innovations.

Because of the inevitable need to make changes to the original concept, Clauses 51 and 52 are included in the Conditions to define the procedures and requirements whereby Variations might be processed. Only the Engineer has the authority to vary the Works by giving an instruction to the Contractor. He is permitted to make changes of almost any kind but generally with the implied purpose of achieving final Completion of the Works.

Clauses 51 and 52

It is obviously necessary for Variations to have some restriction to remain within the intentions of the Parties when a Contract was entered into, although individual Employers and indeed many Engineers contemplate Variations which range from being of little consequence to considerable magnitude. It would be generally accepted that all necessary or desirable Variations are made by the Engineer simply to achieve Completion of the Works as envisaged within the Scope of the Works and no more. They certainly should not commit the Contractor to duplicate the Works or to introduce Works alien to the original concept which would result in a final financial involvement which is out of all proportion to the original Contract Price.

There might be a difference of opinion as to whether certain work constitutes a Variation within the meaning defined under Clause 51, or whether it is already part of the work to be undertaken by the Contractor, but it is stated that the Contractor will not commence any Variation until he has received an instruction from the Engineer. Under Sub-Clause 2.5 all Engineer's instructions are to be in writing but if he considers it desirable, he can give the instruction orally and the Contractor is obliged to comply with this whether or not it is confirmed in writing before obeying the order. If the Engineer gives an oral instruction and does not follow it up in writing,

Clause 51

Sub-Clause 2.5

the Contractor must write to him within seven days confirming the instruction he has been given. If he receives no reply from the Engineer during the next seven days it will be deemed to be an instruction from the Engineer.

The question is sometimes asked about the status of a Drawing issued by the Engineer to the Contractor and whether it might constitute a written Variation. The content of the Drawing will be the predetermining factor in that some Drawings merely amplify or give particular instructions to the Contractor concerning work already covered by the Specification, and obviously these could not be regarded as an ordinary Variation. However, when a Drawing shows a change in the Works as originally contemplated and such a change can be identified as a Variation, the Drawing itself will be accepted as a written order from the Engineer to the Contractor within the meaning of the Clause. If this happens the Contractor must fulfil the requirements of the Drawing in exactly the same way as if he had received a communication in the form of a letter or notice. *Sub-Clause 51.2*

Granting the Engineer the powers to vary the Works is essential, because a Contract without such a Clause would not permit changes of any sort to be made to the physical construction of the Works. This would mean every desired change necessitating a separate or supplementary Contract between the Employer and the Contractor, and that they would have to agree specific rates for the work and consider the effect such work might have on the Contract already in operation.

An instruction in writing from the Engineer is not required for any changes in quantity for work which is merely seen to be the result of the quantities being different from those stated in the Bills of Quantities. It is known that the quantities provided are estimated quantities only for the Works and are not to be taken as the actual or correct quantities to be executed by the Contractor. This therefore permits such changes in quantities to become contractually acceptable without any need for the Contractor to seek a Variation instruction. *Sub-Clause 51.2*

After instructing a Variation, the Engineer must consider how it affects payment to the Contractor and, provided that the work involved is the same in character and within such timing as the work already described in the Bills of Quantities, existing rates for work will be applicable. If the Contract does not contain any rates or prices applicable to the varied work, suitable rates or prices can be agreed between the Engineer and the Contractor after due consultation by the Engineer with the Employer and the Contractor. If there is disagreement, the Engineer fixes such rates and prices he considers appropriate, and until these are agreed or fixed, he decides provisional rates or prices so that interim payments can be made in monthly certificates. *Sub-Clause 52.1*

Sub-Clause 52.1

There are Contracts which appear to have suitable rates for varied work already within the Bills of Quantities, but it must be recognised that such rates in the Bills of Quantities are related to a time-based programme of execution. If the work is done out of the time sequence on which the rates were based, the rates themselves are not applicable. They are then subject to change and must be made reasonable and proper under the prevailing circumstances.

A situation may arise when changes are made which concern a particular *Sub-Clause 52.2* trade or operation, and these changes can influence, directly or indirectly, the reasonableness of rates and prices in the Bills of Quantities for other work which is not directly related to the changed work. If, by reason of a Variation instructed by the Engineer, the rates and prices for this other work are rendered unreasonable or inapplicable, the Engineer, after due consultation with the Employer and the Contractor, will agree suitable rates and prices with the Contractor. As before, in the event of a disagreement, the Engineer fixes rates or prices he believes are appropriate. Provisional rates or prices are used to value interim certificates until rates and prices are agreed or fixed. For example, if Variations affecting the foundations of a bridge caused the construction of the superstructure to be delayed for a considerable period, this Clause would enable the Contractor to seek a change of the rates for ancillary work such as balustrading, kerbing, lighting and surfacing, because of increased costs arising from inflation during the delay period, from working out of sequence, and possible changes in the method of executing the work which would have made the original contract rates unreasonable or unrealistic. Under such circumstances the Contractor is entitled to have these rates altered to a suitable value as may be agreed between the Engineer and himself, even though this would not be a direct part of the Variation itself.

If the Engineer intends to vary a rate or price already in the Bills of Quantities because of the direct effect of the issue of a Variation, or the indirect effect that it might have on other items of work, he must give notice of this intention to the Contractor in writing 14 days after the date of the instruction to make a Variation. No varied work instructed by the Engineer pursuant to Clause 51 shall be valued under Sub-Clauses 52.1 *Sub-Clause 52.2* or 52.2 unless this notice has been given. The Contractor is likewise obliged to give notice in writing to the Engineer of his intention to claim any extra payment or a varied rate or price within 14 days after the date of the instruction.

It is unfortunate that no specified time limit is given within which suitable new rates or prices are to be established. It can be argued that as it is the Engineer's responsibility to measure and value the work fully and at regular intervals in accordance with Clause 60, he would have had to establish the new rates during the same intervals to have fulfilled his obligations to value the work correctly. He can, however, under Sub-Clauses 52.1 and 52.2, use Provisional Rates to enable on-account payments to be included in Certificates issued under Sub-Clause 60.2.

When dealing with the evaluation of Variations, the Engineer would also have to include the direct or indirect effect upon items contained in a preliminary bill, if one existed, and to ensure that such evaluations include the financial effects on such items, not only in respect of any direct cost influence but also of any time factor involved.

The Engineer can only order a Variation in so far as he has the authority to do so, and if he acts outside that authority, the Employer may refuse to pay the Contractor, leaving it to him to seek satisfaction from the Engineer outside the Contract.

The Engineer may order a Variation to omit work but not to have it done by the Employer or another Contractor. Also, he is not permitted to make any Variations during the Defects Liability Period, although he is entitled to require the Contractor to execute any remedial or maintenance work (generally excluding fair wear and tear) during this period, because until all work has been completed the Contractor is responsible for fulfilment of his obligations. Only when this has been done will the Engineer issue a Defects Liability Certificate.

The Engineer need not make a Variation because of a concession or waiver he had made in respect of Specifications, standards, or any changed method of construction which the Contractor may have requested, but he can do so if he wishes. This is an option which he alone can exercise and it would be in the form of an instruction on technical or Specification matters but at no cost to the Employer. If the Contractor varies any work without authority, he is not entitled to be paid for what he does instead of what he should have done according to the Drawings or Specifications. It is arguable whether he should be paid for the work he has done because it will have been a Breach of Contract even if the work is better than that originally required. The Engineer can accept any improved standard of work but need only certify the value of such work at the unit rates and prices in the Bills of Quantities.

Climatic conditions and forces of nature

Some countries appear to experience more predictable weather conditions than others. The seasons are more definable, as are the expected timing of change and duration of hot and cold seasons, dry and wet seasons, hours of light and darkness and high and low wind periods, although recent experience has shown that the unpredictable is occurring more frequently. Other countries have an erratic pattern of weather conditions which seem to defy forecasting. Tendering for work of civil engineering in foreign countries requires the ability to forecast risks associated with climatic conditions which in turn involve additional costs.

The Contractor is not obliged to visit the Site before submitting his *Clause 11.1* Tender, but is deemed to have done so, and to have evaluated for himself the sub-surface, hydrological and climatic conditions and to have made all necessary allowances for them in his Tender Price. He is entitled to *Clause 11.1* be provided with whatever information concerning hydrological and sub-surface conditions has been obtained on behalf of the Employer. If the Employer provides data on climatic conditions this should also become part of the Tender. Therefore, if any such climatic data proves to be wrong, the Contractor might be justified in thinking he has a claim against the Employer.

The risks arising from climatic conditions are basically accepted by the *Sub-Clause 12.2*

Contractor and not by the Employer, but the Contractor can claim additional costs arising from unforeseen physical conditions during the execution of the Works, but not for unforeseen climatic conditions.

Where weather conditions do not conform with expectations the effect on the progress of the Works can be disastrous and give rise to additional costs. Contractors working outside their country of origin and particularly those working overseas for the first time often encounter extreme weather conditions which they have not anticipated. They have little or no experience of their ferocity, intensity, and duration and in particular the influence they consequently have on the Contract programme. Any one of these factors can upset the financial outcome of the Contract to the extent that the Contractor may have to re-examine the Conditions of Contract in much greater depth than he has ever done before.

Clause 44, although not money-making in its own right, can often prove difficult to apply to obtain an extension to the Time for Completion on grounds of bad weather, because it is necessary for the Contractor to prove that the weather conditions causing him the trouble are 'exceptionally adverse'. Considering that it can be difficult to agree about what are 'normal' weather conditions at any given moment, the task of establishing what is 'exceptionally adverse' can be almost impossible. It should therefore *Clause 44* be recognised that 'weather' is one of the by-products of 'climatic conditions' and, according to the actual conditions prevailing, can be classified as a monsoon, typhoon, tornado, cyclone, blizzard, hurricane, storm, tempest etc, and the possible resulting mudslides, avalanches and the like are all the result of climatic conditions. Climatic conditions are referred to in the Contract Documents rather than weather.

If the Contractor encounters any other problem caused by weather or *Sub-Clause 20.4* natural disasters he may seek to recover his additional costs by arguing that it was one of the Employer's Risks whereby he undertook to carry the costs should damage be caused by the forces of nature against which an experienced Contractor could not reasonably have been expected to take precautions. Without knowing exactly what precautions were expected *Clause 21* to have been taken, it is difficult for the Contractor to present a strong argument based on a precise definition, which cannot be found in the Contract, and which will withstand a counter argument. It will be questioned as to whether or not he should have taken some different physical action or if he should have taken some unspecified financial action. The former is a matter of technical opinion whereas the latter may be one of the need for insurance.

Likewise, without a precise definition of the forces of nature in the *Sub-Clause 20.4* Contract documents, it is only possible to assume what these may be. If accepting that climatic conditions can be taken as the effect of the weather, could it not be assumed that these forces of nature are intended to refer to earthquakes, volcanic eruptions, tidal waves, landslides and suchlike, each being the result of natural forces but not necessarily part of the climatic conditions? Difficulties can arise, of course, and the Parties and the Engineer will need to agree about cause and effect. For example, flooding on Site can be caused in a number of ways: excessive rainfall due to unexpected climatic conditions; rivers overflowing due to normal rainfall or blockages upstream of the Site; man-made causes such as the collapse of a dam or bridge.

If the progress of the Works is suspended because of climatic conditions, the Contractor is not entitled to expect a notice from the Engineer instructing him what to do and a promise of payment when the work has been carried out. The Contractor has to do whatever is necessary to secure the works at his own cost, although he should be granted an extension of the Time for Completion if exceptionally adverse climatic conditions are causing the problem. Securing extra payment for such a delay is a different matter and requires action under a different Clause depending on the actual circumstances.

Sub-Clause 44.1

There remains the particular situation where the Contractor is required to carry out certain Works arising from Variation Orders because of instructions from the Engineer or actions of the Employer. If such work has to be done during periods of adverse climatic conditions and outside his Programme of Works, he may be entitled to an adjustment of the existing Bill of Quantities rates or to new rates where necessary, and to additions to the Contract Price to cover additional time and overheads. Possibly the best advice to a Contractor is for him to make certain that if any possibility exists of him incurring additional and unrecoverable expense through weather, climatic conditions or even the forces of nature, he should cover this in his Tender Price.

Clauses 51 and 52

Unless a strong case can be made by the Contractor that the programme of operations has been delayed by reasons beyond his control, he has little chance of success with a claim. Reasons he may put forward could be that many instructions or Variation Orders issued by the Engineer caused disruption or a change to the timing of various operations, and the possibility of suspension or other factors that would turn what was expected to be summer working into winter working.

Fossils

The discovery of fossils, coins, articles of value or antiquity, structures etc, on the Site usually causes more of an inconvenience to the Contractor than major physical obstructions. Items of antiquity can be a cause of concern for the Contractor be they as small as an ancient bracelet or as large as the remains of buildings or settlements of a bygone age or civilisation, and the problems arising can be very significant in relation to the value of the object or structure uncovered. In the event of such a discovery, the Contractor must tell the Engineer and carry out his instructions for dealing with the situation. If the Contractor suffers delay or incurs extra cost by following the Engineer's instructions, an Extension of Time and an addition to the Contract Price will follow after due consultation by the Engineer with the Employer and the Contractor.

Clause 27.1

The fossils Clause does not state that any extra work will be dealt with as a Variation Order, which would guarantee a profit margin to the Contractor on the cost of such work, but only that he will recover the costs incurred which, according to the definition in Sub-Clause 1.1(g)(i), can include an allowance for overheads alone. This Clause only refers to an Engineer's instruction for 'dealing with the same', and not to other indirect but often necessary operations which must go on if the Works are to be completed on time. What happens to the rates in the Bills of Quantities when they become unreasonable because of the effect of these particular circumstances? Can they be changed in the same manner that would apply if the additional work was regarded as a Variation Order, or are the side effects of 'dealing with the matter' to be at the expense of the Contractor? Although there is no positive direction as to what the Contractor must do in the event of a fossil problem, it might be argued that, as it is the Engineer who must give instructions as to what he requires to be done, the consequences of what was actually done following these instructions could be treated as being a Variation and the Contractor paid accordingly under Clauses 51 and 52. *Clause 27.1*

Clauses 51 and 52

Of course all discoveries are not small fossils. Most problems arising under this Clause are related to the discovery of geological or archaeological matters rather than just fossils, coins, or articles of value or antiquity. The immediate effects, as well as the long term ones, arising from such circumstances can be considerable changes to the cost of the Works as well as to the time required to complete them. If the discovery is large or of major importance, the influence on the Contractor's anticipated cost of the Works could be significant and could include some of the following

(a) dislocation to his Programme of Work
(b) possibility of a suspension, which could be dealt with as a Suspension Order under Clause 40
(c) loss of production and inability to obtain maximum productivity from labour plant and other resources if the work is not suspended
(d) extension to the Time for Completion which, although mentioned as such in Clause 27, will allow for the recovery of additional costs
(e) extra cost of management, staffing and certain other overheads and on-costs; most of these should be recovered
(f) increased cost of Bonds, Insurances, Contract funding and such like
(g) processing a claim under these circumstances through Clause 53.

The Contractor should remember that any items discovered belong to the Employer. *Clause 27*

Suspension of the Works

Clause 40 and Sub-Clause 69.4

Under the third edition of the FIDIC Conditions the Engineer alone is entitled to suspend the progress of the Works as laid down in Clause 40. Clause 40 in the fourth edition is similar, but it also contains a Sub-Clause,

40.2, which requires the Engineer to consult the Employer and the Contractor before determining any Extension of Time or any money to be awarded to the Contractor. Clause 40 of the third edition required the Contractor to give notice of his intention to claim time and money arising from suspension, and the Engineer to settle the claim and determine the amount due. Under the fourth edition the Engineer takes the initiative to determine any time and money due to the Contractor.

The fourth edition now provides, as set out in Sub-Clause 69.4, for the Contractor to be entitled to suspend work or reduce the rate of work if the Employer is in default and does not pay him the amount due under any certificate within 28 days of the time allowed for the Employer to make payment (Sub-Clause 60.10). This is a new concept that gives the Contractor a means of putting pressure on the Employer to pay on time, and without having to resort to terminating his employment, as was the case in the third edition.

In Sub-Clause 40.1 of the fourth edition, the Engineer must give *Sub-Clause 40.1* instructions to the Contractor to suspend the progress of the Works or of any part for such time and in such a manner as he considers necessary. During Suspension of the Works the Contractor has to protect and secure the Works or any part affected to the extent required by the Engineer. If the suspension has arisen for any of the reasons defined in Sub-Clause 40.1, the Contractor must bear the cost of suspending the Works or any part and the possibility of losing construction time. The reasons are that

(*a*) the Suspension of the Works was already provided by the Contract or within the Contractor's programme, in which case the costs of such suspension were deemed to be in the Contract Price and the time factor within the Time for Completion
(*b*) the Contractor was in default or in Breach of Contract
(*c*) the suspension was required because of climatic conditions on Site. This can cause complications because under Sub-Clause 20.4(h), the Employer accepts the risk of loss or damage caused by any operation of the forces of nature against which an experienced Contractor could not reasonably have been expected to take precautions. No doubt such an occurrence of damage caused by the forces of nature would be raised by either the Engineer or the Contractor and action taken as laid down in Sub-Clause 20.3
(*d*) the suspension arose to ensure the proper execution of the Works or for the safety of the Works or for any Part of the Works.

When the suspension is brought about by any conditions other than those defined in Sub-Clause 40.1, the Engineer consults the Employer and the Contractor and determines the Extension of Time due to the Contractor under Clause 44 and the amount to be added to the Contract Price. Sub-Clause 40.3 covers the procedures required to bring the suspension to an end should the Engineer fail to instruct the Contractor to resume work within 84 days of the date of suspension, in cases where the suspension is not due to the causes set out in Sub-Clauses 40.1(a), (b), (c) or (d).

An end to any suspension arising from the causes listed in Clause 40.1 will be provided by the circumstances described and by the Engineer if

(a) under the Contract or programme there is a date specified for the resumption of work
(b) the Contractor rectifies any default as quickly as possible to save time and money; otherwise he risks having his employment terminated under Clause 63
(c) by nature if conditions were adverse
(d) the Contractor in his own interests must resume work as soon as possible to persuade the Engineer to lift the Suspension Order.

Apart from these circumstances, the Contractor has the right to give notice to the Engineer requiring permission within 28 days of giving notice to resume work. If no permission is given within these 28 days, the *Clause 51* Contractor has the right, if the suspension covers only a part of the Works, to deem such part to have been omitted under Clause 51, or if the suspension covers the whole of the Works, to treat it as an event of default by the *Clause 69* Employer and terminate his employment in accordance with Clause 69.

With regard to the Contractor's right to suspend work or slow down the rate of work in the event of late payment by the Employer, Sub-Clauses *Sub-Clauses 69.4* 69.4 and 69.5 cover the resumption of work or normal working. If the *and 69.5* Contractor is not paid the amount due under any certificate, subject to any authentic deductions by the Employer, within 28 days of the expiry of the time stated in Sub-Clause 60.10, the Contractor is entitled to suspend work or reduce the rate of work 28 days after the giving of notice to the Employer, with a copy to the Engineer. If the Contractor suffers delay or incurs cost because of this action, the Engineer has to consult the Employer and the Contractor before determining any Extention of Time or addition to the Contract Price and has to notify the Contractor and Employer accordingly.

Clause 69.4 is clear as to the extent that the Contractor is entitled to an extension of time and costs resulting from the suspension of work or the reduction of the rate of progress, but, as most Contractors are well aware, the effect of such an interruption to their programme can only mean the beginning of a far wider involvement of time and cost than just for the period of interruption itself.

When the Contractor has been paid the money due because of the suspension or reduced rate of working, with interest pursuant to Sub-Clause *Sub-Clause 60.10* 60.10, and, if notice of termination has not been given by the Contractor under Sub-Clause 69.1, he must resume normal working as quickly as possible.

The introduction in the fourth edition of this right of the Contractor to suspend work or reduce the rate of working will no doubt cause difficulty to both the Contractor and the Employer, but more particularly to the Engineer who will have the unenviable task of translating the temporary stoppage or reduction in the rate of progress into acceptable values of time and money.

Defective work

One of the most important certificates sought after by a Contractor must be the Defects Liability Certificate because it is the Engineer's confirmation to the Employer that the Contractor has completed all the Works and remedied all the defects to the Engineer's satisfaction. This certificate has to be issued within 28 days after the expiry of the number of days stated as being the Defects Liability Period in the Appendix to the Tender, or, if the Works have been done in sections or as parts of the Permanent Works, is to be issued after the expiration of the Defects Liability Period of the last Section or Part to be completed. If, however, the Engineer has required the Contractor to undertake works of amendment, reconstruction, remedying of defects, shrinkages or other faults, he will only issue his Defects Liability Certificate when these have been completed, especially if such completion was after the expiry of the Defects Liability Period.

Sub-Clauses 61.1 and 62.1

So what then are defects and how are they to be understood within the meaning of the Conditions of Contract? Can defects exist during the construction of the Works or do they only come into being once the Works have been substantially completed? It is common practice during the construction period to refer to any work which is not as it should be as 'defective work'. This may be correct as a matter of technical definition, but it is not correct as a proper contractual definition. Work which is not correctly executed becomes simply work not conforming to the Specification during the Construction Period.

The Contractor has to rectify sub-standard work if he wants to be paid for it. It is not the Engineer's duty to point it out to the Contractor or to instruct him what to do. All the Engineer must do is ensure that any Certificate of Payment, when issued, does not include payment for any work until it has been done properly and to his satisfaction.

Sub-Clause 60.4

Defective work or defects only come into proper prominence around the time the Contractor makes a request to the Engineer for a Taking-Over Certificate. This is when he considers that the Works are substantially completed, even if there still remain certain works incomplete but which could be finished during the Defects Liability Period. Such an arrangement can be accepted quite properly by the Engineer provided that the Contractor gives a written undertaking when he makes his request for a Taking-Over Certificate that he will complete such works during the Defects Liability Period.

Before the Engineer issues a Taking-Over Certificate he will carry out an inspection and if he is satisfied he will issue the certificate to the Employer with a copy to the Contractor within twenty one days of receiving the notice from the Contractor. If there remains work to be done before he can issue his Taking-Over Certificate he will specify to the Contractor exactly what is required. Only after this work has been done will a Certificate be issued. This has to be within a period of twenty one days after the satisfactory completion of the specified works.

Between the giving of notice to the Contractor of the outstanding works and the completion of these works, the Engineer can notify the Contractor of any defects in the Works which he requires to be rectified before he issues the Taking-Over Certificate. Only after the completion of the oustanding works and the remedying of defects can the Engineer issue the certificate.

During the Defects Liability Period, the Contractor is required to complete any works listed and not approved as being outstanding on the date stated in the Taking-Over Certificate and to execute all such works of amendment, reconstruction, remedying of defects, shrinkages or other faults as the Engineer may instruct.

Sub-Clause 49.2(b)

The responsibility of the costs incurred carrying out repairs and such like depends entirely on the Engineer's opinion. If the defects arose through his own or the Employer's fault, the Contractor will be paid for the rectification. If it is the Contractor's fault he will not be paid. In all of this the Contractor may undertake any work which needs to be done even if caused by fair wear and tear, and any work which has arisen through causes outside the Contractor's control will be paid for by the Employer in accordance with Clause 52.

Sub-Clause 49.3

Defects are therefore only those works which are found and notified as being unacceptable after the Engineer's instructions to the Contractor about work to be done before the issue of the Taking-Over Certificate, or work discovered after the issue of a Taking-Over Certificate. If the same unacceptable work was in existence before such a certificate, it was simply items of work remaining to be carried out properly by the Contractor to the satisfaction of the Engineer. Imperfections in the Works are not all caused by the Contractor. They could be the fault of the design of the Works by the Engineer or the Employer's designers or because of a faulty Specification. Again, such imperfections are not 'defects' whether discovered before or after the Taking-Over Certificate, but they are required to be given an identity of their own more appropriate to the situation under which they occur. In such instances this would be the same as damage arising from a risk accepted by the Employer and for which the Contractor is entitled to recover the costs of rectification if ordered to do so.

5. Claims — new procedures

The mental gynmastics performed in attempts to convince the Engineer that the Contractor has suffered extensive losses due to a variety of circumstances and events, the responsibility for which the Engineer is usually accused of, or due to circumstances arising for which no allowance was made in the Tender Price, are well known to Engineers and Contractors alike. The Contractor often ignores the fact that the problem was caused by him or was his responsibility under the Contract. More journalistic endeavours have been devoted to the subject of claims than any other, and most publications dealing with claims are read eagerly by Contractors. Claims will always arise because the Conditions make provision for their proper presentation and also contain the procedure by which they will be dealt with by the Engineer. Considering the nature of the fourth edition, of the environment and of the human race, it is a rare event for a major Contract to be completed on time, within budget and without involving any Variations, errors, faults or unpredicted forces of nature. All claims submitted by the Contractor should be cogent, logically argued, well prepared and presented to receive proper consideration by the Engineer — he will become hostile and angry if he has to read through a lot of rubbish, misquoted facts and what is obviously a blatant attempt to secure extra payment and time when they are not due.

Presentation of claims: Clause 53

The new Clause 53 gives positive guidelines to achieving success in the preparation and presentation of claims. The most important is that the Contractor gives proper and timely notice to the Engineer. It will be helpful for the Contractor to send copies of all claim documents submitted to the Engineer to the Engineer's Representative as well. This should ensure immediate observance of the facts related to the events on Site that give rise to the claim. The Clause requires copies to be sent to the Employer in every case. It is no good for the Contractor to submit his claims to the wrong person, to the wrong place or outside the permitted time limits allowed. The Engineer's address is given in Part II and the Engineer's Representative's address is generally that of his Site Office. *Clause 53*

Any claims which the Contractor wishes to present must be made before he submits to the Engineer a Statement at Completion, which itself must *Sub-Clause 60.5* be presented no later than 84 days after the issue of the Taking-Over Certificate. Because other claims can arise after the date of such a certificate, these additional claims must be contained within the Final Statement. Failure to include them releases the Employer from any liability to even consider them. However, the Contractor can reduce the risk of failure with his claims if he makes a point of including all of them with his monthly Application for Payment even if at the time he is unable to place a detailed valuation on his claim.

Sub-Clause 60.6

The Engineer has a duty to the Employer to accept only those claims which are valid under the Conditions of Contract — there is no such thing as an 'extra-contractual' claim. Either the claim exists under the Conidtions of Contract, or it does not. Despite this, this type of claim has been and is still used by Contractors. It is used to bring to the attention of the Engineer a request for extra payment when no suitable Clause to provide support for such a payment exists in the Conditions of Contract.

It might seem a waste of time and effort to submit such an unsupportable request, but it does serve a useful purpose. It introduces a 'bargaining factor' for use in future negotiations concerning payment. Although it does not become a proper part of the Final Certificate of the Works, it may get included as an *ex gratia* payment without the admission of liability. This *ex gratia* payment can often avoid potential litigation or Arbitration proceedings, the costs of which could be many times greater than any payment otherwise made. It must always be remembered that an *ex gratia* payment made without the admission of liability provides no support whatsoever to the argument that the Party making it is admitting any guilt.

Contractors make claims in many ways. The following list indicates the range of items which often form the subject of claims. It is random and not intended to cover all possibilities.

Action by Employer
Actions of a Nominated Sub-Contractor
Adverse weather conditions
Application of rates in the Bills of Quantities
Awaiting Drawings and instructions
Delay during the execution of the Works
Delay in being given approval by the Employer or Engineer
Delay in payment
Delay in the Commencement of the Works
Difficulties with Customs
Difficulties with suppliers
Disputes over quantities
Disruption, impact of disruption and extended overheads
Employer's Risks
Errors in setting out arising from incorrect data from the Engineer
Exploration
Fossils, antiquities etc.
High number of Variation Orders

Interpretation of Specification
Local customs
Method of construction
New items or rates in the Bills of Quantities
Physical conditions
Possession of and access to the Site
Special risks
Strikes
Substitution of materials
Suspension
Tests
Under-utilisation of resources
Waivers
Work permits

The actual presentation of a claim can be made in a variety of ways and many Contractors prefer their own particular form. The following suggests some approaches.

When presenting a claim under Sub-Clause 53.3, it is advisable to summarise the subject matter being reviewed first to save the Engineer and the Employer having to read through a number of pages without appreciating the precise purpose of the claim until the end. It is better to start with an introduction which gives a brief outline of the Contractor's submission and to develop the claim in detail later.

Sub-Clause 53.3

Historical background

It is helpful to give an early resumé, in as much detail as practicable, of any historical data affecting the subject matter of the claim and to make all necessary references to other documents to enable the reader to appreciate the background.

Contractual argument

This is an important part of the presentation. It should state the particular Clause or Clauses on which the claim is founded, and then the Contractor should set out in detail a logical argument intended to bring the Engineer as close as possible to the Contractor's understanding of the claim. It may be possible to refer to similar known claims which have been successful in either litigation or Arbitration in the country in which the Project is being constructed or according to the law of the country defined in the Contract.

General

There are two very important matters to which the Contractor must direct his attention before any claim has a chance of succeeding. First, from the beginning of the Contract he should ensure that the actual cost he incurs in carrying out the work is recorded from the level of site activities upwards in such detail as to permit each technical and administrative operation to be evaluated separately wherever possible. Second, he must ensure that a Site Diary is kept on a daily basis and that proper labour and plant records are maintained, and if possible, for them to be agreed with the Engineer's Representatives as being statements of fact.

These two factors are important because the evaluation of all claims is made from 'contemporary records' and not necessarily from existing rates in the Bills of Quantities. Exactly what is meant by contemporary records is a matter to be decided by the Engineer, but their origin could be related, among other things, to those actual costing records maintained by the Contractor. The contemporary records relating to any claim start from the actual timing of the 'event' rather than to the giving of notice, thus indicating that the Contractor needs to keep proper and suitable cost records available from the beginning of the Contract to be able to identify items which support the claims.

Sub-Clause 53.2

Supporting data

Matters dealing with this particular part of the Contractor's claim can be recorded under various subject headings and presented as an appendix if desirable. This data may include items, wherever possible, such as: Site records; photographs; Site Diaries; daily weather reports; charts and maps; instructions; quality control documents; programmes; Drawings (dated when received); correspondence; daily weather records; Tender analysis; analysis of appropriate unit rates; invoices; wage sheets; plant; fuel and labour records; minutes of meetings; visitors; and any other matters supporting the claim.

Financial

It must be appreciated that there is no set format or standard presentation of financial analysis which will be applicable to each and every claim. The financial presentation is made to suit the prevailing circumstances, but the structure of most financial analyses is basically a detailed comparison between the costs anticipated by the Contractor when tendering compared with the costs incurred when actually doing the work, and being affected by the circumstances which gave rise to the claim, and any consequential costs.

Application of Clause 53

To understand more clearly the actual route to follow in giving notice and presenting a claim, Chart 4 sets out the basic actions to be taken according to *Clause 53* and the timing sequences involved. These actions are as follows.

(*a*) Start contemporary records simultaneously with occurrence of 'event'.   *Sub-Clause 53.2*

(*b*) Give notice of intention to claim within 28 days of the event.   *Sub-Clause 53.1*

(*c*) Act on the Engineer's instructions about the additional records.   *Sub-Clause 53.2*

(d) Substantiate claim for a single claim within 28 days of giving notice (possibly varied by agreement with the Engineer).   *Sub-Clause 53.3*

(e) For a continuing claim, submit detailed particulars on a regular basis (intervals to be agreed with the Engineer).   *Sub-Clause 53.3*

(f) For this continuing claim, send a Final Account within 28 days of the end of effects resulting from the event.   *Sub-Clause 53.3*

(g) Include an Application for Payment of the claim in each application for Certificate of Payment in addition to actions required under Sub-Clause 53.3.   *Sub-Clause 53.3*

(h) Take heed of Clauses 60.5, 60.6 and 60.9 which limit the time for making claims.   *Sub-Clauses 60.5, 60.6 and 60.9*

Claims Clauses

There are certain Clauses which provide the Contractor with opportunities for increasing the Contract Price and certain Clauses which entitle the Employer to receive payment from the Contractor. These Clauses are listed in Tables 1 and 2, and indicate whether extra costs (C), extra time (T), and profit (P), are applicable.

Table 1. Clauses of Contractor's extra payments

Clause number	Clause title	Adjustment*
5.2	Ambiguities (depending on circumstances)	T+C
6.3 and 6.4	Engineering Drawings delay	T+C
12.2	Physical conditions	T+C
17.1	Setting out (errors based on incorrect data)	C+P
18.1	Exploratory boreholes	C+P
20.3	Repairs and Employer's Risks	C+P
27.1	Fossils, antiquities, structures	T+C
31.2	Opportunities to other Contractors	C+P
36.5	Tests	T+C
38.2	Uncovered work	C
40.2	Suspension	T+C
42.2	Employer's failure to give possession	T+C
49.3	Cost of remedying defects	C+P
50.1	Search for defects	C
52.1	Variations	C+P
52.1 and 52.2	Extra payment for Variation Orders	C+P
52.3	Fifteen per cent reduction or increase	±C
65.3	Damage to Works by Special Risks	C+P
65.5	Increased costs arising from Special Risks	C
65.8	Termination of Contract	C and C+P
69	Defaults by Employer	T+C
70.1	Increase or decrease of cost	by formula
70.2	Changes in legislation	±C
71	Currency and Rates of Exchange	C+P

*T = time adjustment; C = cost adjustment; P = profit adjustment.

Time limitation

For a claim to be accepted even for consideration, the correct observation *Sub-Clause 60*
has to be given to 'time limits' within which certain formalities have to
be met.

Statement of Completion

The Contractor has to submit to the Engineer a Statement of Completion *Sub-Clause 60.5*
not later than 84 days after the issue of the Taking-Over Certificate for
the whole of the Works, including amounts or estimated amounts of all
claims.

Draft Final Statement

The Contractor has to submit to the Engineer a Draft Final Statement *Sub-Clause 60.6*
not later than 56 days after the issue of the Defects Liability Certificate
covering the value of work done and any further sums due.

Claims not included

The Employer is not liable to the Contractor for any claims not included *Sub-Clause 60.6*
in the Final Statement or Completion Certificate. This is clearly set out *Sub-Clause 60.5*
as part of the Cessation of Employer's Liability. Many claims will be settled *Sub-Clause 60.9*
during the construction period before Taking Over occurs. Some will be
settled during the Defects Liability Period, and some may be resolved during *Sub-Clause 60.8*
the 28 days between the submission of the Final Statement by the Contractor
and the issuing of the Final Certificate by the Engineer.

Settlement of disputes

The Contractor can use Clause 67 at any time to precipitate the settlement *Clause 67*
of any unresolved claim. Equally, the Employer can dispute the Engineer's
decision and also use Clause 67 to seek reduction of any claim determined
by the Engineer.

Table 2. Clauses for Employer's recovery of money from Contractor

Clause number	Clause title	Notice*
25.3	Contractor's failure on insurance	NN
30.3 and 30.4	Damage to highways and bridges	Consult
37.4	Rejection of materials and plant	EN
39.2	Contractor's failure to obey Engineer (improper work and materials)	EN
46.1	Rate of progress	EN
47.1	Failure to complete on time (liquidated damages)	NN
49.4	Failure of Contractor to do repairs	EN
59.5	Failure to prove payments to Sub-Contractors	EN
63.3	Default by the Contractor	EN
64.1	Urgent remedial work	EN
65.8	Payment on Termination	EN

*EN = Engineer's notice; NN = no notice.

Conclusion

The Contractor must understand quite clearly under the fourth edition Conditions of Contract that to have a chance of succeeding with any claim he must follow precisely the procedures laid down in Clauses 53 and 60.

Clauses 53 and 60

Claims and entitlements

Despite the emphasis placed on the application of Clause 53 in connection with claims, there exists a school of thought which considers this to be incorrect, that claims can only exist when the Contractor disagrees with a 'determination' made by the Engineer, and that where such determination is accepted by the Contractor without argument, it is the end of the matter and that whatever has been determined by the Engineer does not need further processing through Clause 53. Thereafter it finds its natural way into the system of evaluation and future payment.

Clause 53

The background to this argument is that there is a positive difference between a claim and an entitlement. A claim is a matter which the Contractor, not the Engineer, instigates, and is to be duly processed through Clause 53. An entitlement does not necessarily depend on the Contractor taking any specific action to instigate proceedings or the giving of notice, because there are many Clauses specifying a duty for the Engineer to have due consultation with the Employer and the Contractor and then for him alone to determine any extra cost and time due to the Contractor. All of this is considered a straightforward matter of routine to be performed by the Engineer, with the Contractor being advised only of the outcome.

It may be of interest that the same school of thought acknowledges that a claim situation can arise once the Engineer has made a determination, but which the Contractor finds unsatisfactory. In such a case the Contractor is faced with two options: to recognise that a determination has been given by the Engineer, and then to make a claim under Clause 53 for the difference between the amount as determined and such an amount as the Contractor considers to be reasonable and proper; or to dispute the amount determined by the Engineer and to process it through Clause 67 and refer it to Arbitration.

Clause 67

Readers can only form their own opinions as to which view is correct until the fourth edition is in operation and such situations are tested in the many legal systems. The Authors strongly recommend that whenever the Contractor is seeking means to increase the Contract Price under any particular Clause, then even where the Engineer is obliged to make a determination, that the Contractor applies the full requirements of Clause 53 and gives all proper notices and record data as required, both under this Clause and other Clauses with the purpose of receiving additional payment and Extensions of Time, and to be paid accordingly.

6. Charts of selected Clauses

Things would be much more difficult if Contract documents relied solely on the written word and there were no Drawings, maps, programmes or graphics to illustrate exactly what is required. Considerable attention is always given to the written words which form the Contract documents, and in particular, the actual Conditions of Contract. These Conditions can vary in presentation depending on who was responsible for the actual drafting, preparation and assembly. If this was done by a legally-minded person, the Conditions will comprise a document of considerable detail and accuracy. If it was prepared by someone more technically-minded, it can be a document full of detail but often miss important definitions as to who is responsible, who is to pay for what is required, and when.

The new fourth edition of FIDIC clearly denotes the obligations and responsibilities of those involved with the Contract and sets out the functions, purpose and operation of each Clause in detail. However, it is sometimes difficult for the reader to understand and remember the exact content of a Clause and what it is intended to convey to the Parties, particularly with so many options of timing and differences of wording. The following Charts are intended to clarify the various situations covered by a Clause.

Clauses are codified as follows

(a) nodes are commencement and termination of times allowed for certain requirements to be observed
(b) serrated lines represent the time allowance between the nodes
(c) straight lines act as connections between one operation and another.

With this basic information it is possible for every Clause that presents a difficulty of immediate understanding to become clear as to its purpose, the particular requirements to be observed therein and exactly how each operation and function contained are related to each other and how they can be processed. Proficiency in preparing such charts will come with practice. Example Clauses to show the application of the charts are

(a) Clauses 41, 42, 43 and 48. Commencement, Extension of Time and Completion (chart 1)
(b) Clause 12. Adverse Physical Conditions (chart 2)

Text continues on p. 83

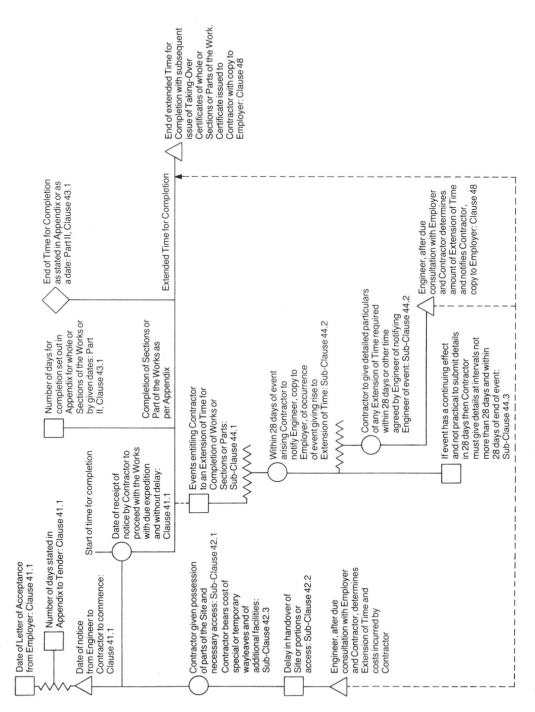

Chart 1. Clauses 41, 42, 43 and 44: Commencement, Extension of Time and Completion

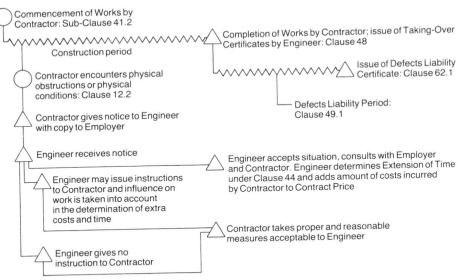

Chart 2. *Sub-Clause 12.2: Adverse Physical Conditions*

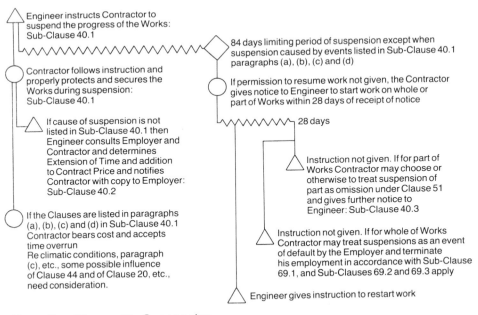

Chart 3. *Clause 40: Suspension*

Continued from p. 80

(c) Clause 40. Suspension (chart 3)

(d) Clause 53. Procedure for Claims (chart 4)

(e) Clause 67. Settlement of Disputes (chart 5)

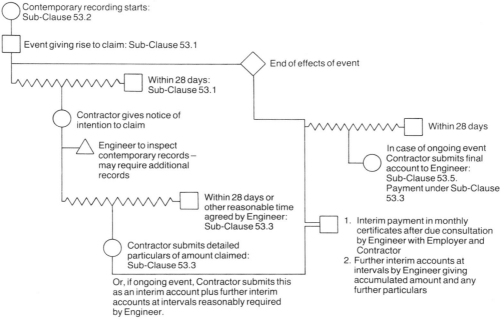

Contemporary recording starts:
Sub-Clause 53.2

Event giving rise to claim: Sub-Clause 53.1

End of effects of event

Within 28 days:
Sub-Clause 53.1

Contractor gives notice of
intention to claim

Engineer to inspect
contemporary records –
may require additional
records

Within 28 days

In case of ongoing event
Contractor submits final
account to Engineer:
Sub-Clause 53.5.
Payment under Sub-Clause
53.3

Within 28 days or
other reasonable time
agreed by Engineer:
Sub-Clause 53.3

Contractor submits detailed
particulars of amount claimed:
Sub-Clause 53.3

1. Interim payment in monthly
certificates after due consultation
by Engineer with Employer and
Contractor
2. Further interim accounts at
intervals by Engineer giving
accumulated amount and any
further particulars

Or, if ongoing event, Contractor submits this
as an interim account plus further interim
accounts at intervals reasonably required
by Engineer.

Notes
1. Contemporary records form basis of payments – no records,
no payment: Sub-Clause 53.4
2. Engineer determines payment only on particulars as submitted:
Sub-Clause 53.5
3. Claims must conform with time limits referred to in Clause 60

Chart 4:   Clause 53: Procedure for Claims

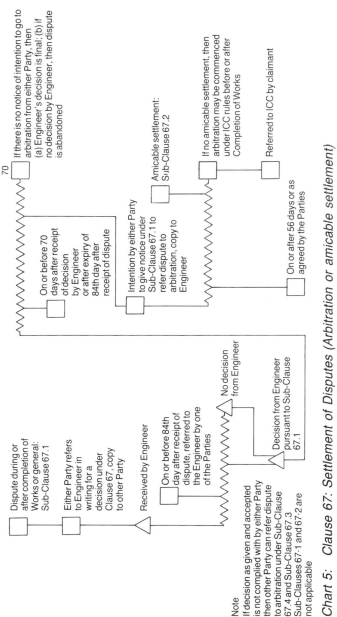

Chart 5: Clause 67: Settlement of Disputes (Arbitration or amicable settlement)

Dispute during or after completion of Works or general: Sub-Clause 67.1

Either Party refers to Engineer in writing for a decision under Clause 67, copy to other Party

Received by Engineer

On or before 84th day after receipt of dispute, referred to the Engineer by one of the Parties

Decision from Engineer pursuant to Sub-Clause 67.1

No decision from Engineer

Note
If decision as given and accepted is not complied with by either Party then other Party can refer dispute to arbitration under Sub-Clause 67.4 and Sub-Clauses 67.3 Sub-Clauses 67.1 and 67.2 are not applicable

On or before 70 days after receipt of decision by Engineer or after expiry of 84th day after receipt of dispute

70

If there is no notice of intention to go to arbitration from either Party, then (a) Engineer's decision is final; (b) if no decision by Engineer, then dispute is abandoned

Intention by either Party to give notice under Sub-Clause 67.1 to refer dispute to arbitration, copy to Engineer

Amicable settlement: Sub-Clause 67.2

On or after 56 days or as agreed by the Parties

If no amicable settlement, then arbitration may be commenced under ICC rules before or after Completion of Works

Referred to ICC by claimant

7. Arbitration and disputes

During the course of a Contract many differences of opinion can occur between all the Parties, probably of a technical nature during the construction period, but towards completion they are more likely to be about the value being placed on the work performed. This is because it is only towards the end of the construction period that it is possible to have a reasonably clear indication of the probable value of the Final Certificate, and it is then when differences of a financial nature take on more importance.

Because differences of a technical and financial nature can and do occur, there exists within the Contract the facility for referring differences which cannot be resolved to an independent authority who will decide objectively, on the basis of the evidence and arguments submitted, who is right and who is wrong, and how any award will be implemented. Arbitration is, and will remain so for some time, one way of settling differences in a reasonably peaceful and civilised manner. Although the rules of Arbitration may differ in various parts of the world, those selected by FIDIC are those of the International Chamber of Commerce (ICC) in Paris, France, whose rules apply under this Contract. Where an alternative set of rules is preferred by the Employer, provision is made in Part II. Although the ultimate Settlement of Disputes under this particular form of Contract is only by Arbitration there are other recognised means whereby disputes can be resolved. These are conciliation, mediation, litigation and compromise.

Clause 67

Sub-Clause 67.3

Conciliation

The means whereby a referee is appointed who will listen to the complaints and reasons of rejection made by those in dispute and seek to persuade the Parties to reach an acceptable solution without resorting to Arbitration.

Mediation

Where a third Party listens to the respective arguments of those involved and forms his own opinion as to what is the correct solution, and then seeks to persuade both Parties to accept his views and settle their dispute.

Litigation

When those involved seek a settlement in a Court of Law because the matter in dispute is more orientated towards requiring a legal ruling on the wording of the Contract rather than on matters of technical differences.

However, when a Contract contains a Clause setting out that differences and disputes are to be settled by Arbitration, attempts to enter into litigation before applying this Clause may be unfruitful and prejudice the Arbitration. Indeed, litigation may be resisted and the right to Arbitration demanded by either Party to the dispute.

Compromise

Nowadays, when the Employers are large conglomerates and public bodies, it does happen that, rightly or wrongly, they exert market pressures on the Contractor and the Engineer to reach a compromise settlement to the dispute in question. Although injustices can happen, a *quid pro quo* arrangement can often become acceptable when the comparative costs of reaching a settlement by this means are more favourable.

Arbitration

With particular reference to the fourth edition of the FIDIC Conditions of Contract, the differences or disputes once identified need not wait until the Works have been completed, but can be referred to Arbitration at any time. It must be recognised by both Parties, however, that their *Clause 67* responsibilities and obligations under the terms of the Contract are in no way altered. The right of the Parties to refer disputes to Arbitration is already predetermined, by the Conditions of Contract, to be under the ICC Rules of Conciliation and Arbitration. However, Part II Clause 67 may allow a different Settlement of Dispute procedure, other than that of the ICC, and that Clause 67 is varied accordingly. It is interesting that the fourth edition of the Conditions of Contract refers to the need for the Parties to seek an 'amicable settlement' as a prerequisite to the actual commencement of Arbitration — this was not required in the third edition.

The wording of Sub-Clause 67.2, in particular the second sentence, *Sub-Clause 67.2* 'Provided that, unless the parties otherwise agree Arbitration may be commenced . . .', suggests that the Parties have at least three options to pursue. First, they may agree to waive the requirements to attempt to reach an amicable settlement in which case Arbitration proceedings could commence as soon as the claimant Party gives notice to the other that the dispute is to be referred to Arbitration. Secondly, the period of 56 days for reaching an amicable settlement might be reduced, and third, the period might be increased, either way, to suit the prevailing circumstances.

According to the present ICC Rules, the Settlement of Disputes will be made by one or more Arbitrators appointed in accordance with such rules, and who will have full powers to open up and revise any decision, opinion, direction or valuation of the Engineer. Neither party is limited in any way to submitting evidence or arguments to the Arbitrator already put before the Engineer for the purpose of obtaining his decision in the first place under Clause 67. Although the ICC will be acting under its *Sub-Clause 5.1(a)* own particular rules and procedures, it is important to recognise that unless these rules conform to the Statutes or legislation operating within the law stated in Sub-Clause 5.1(a), the Laws of the country will always take precedence. This is understandable because if an Arbitration award is to have the legal support of the country in which the Works are being constructed, then unless these Arbitration procedures conform to these laws and requirements, an award made might not be enforceable in that country. The ICC *Guide to Arbitration*[4] covers the Applicable Law under Item 4.8 in the 1983 issue as follows.

(*a*) The rules applicable to the proceedings are distinct from the law applicable to the merits of the case.

(*b*) To ensure that the award will be enforceable in law, the mandatory rules of national law applicable to International Arbitrations in the country where the Arbitration takes place must be observed even if other rules of procedure are chosen by the Parties or by the Arbitrator.

(*c*) The Parties are free to choose the rules of the proceedings. These may consist of the national procedural law where the Arbitration takes place, a procedural law other than that of the country where the Arbitration takes place, or no national law at all, but merely the rules that the Parties agree to. If the Parties fail to specify the applicable rules, the Arbitrator will decide them, remembering that they must be compatible with the ICC Rules.

(*d*) When the Arbitrator determines the procedural rules himself, he may either derive them from a national law or draw up his own.

(*e*) The choice of this law (the law applicable to the substance of the dispute) is also left entirely to the Parties. In the absence of an agreement between the Parties on the point, the Arbitrator decides which law is applicable.

(*f*) When making the choice, the Arbitrator takes into account the most appropriate law in view of the circumstances of the dispute.

(*g*) As required by the ICC Rules, the Arbitrator must always remember the provisions of the Contract and relevant trade usages.

Referral of a dispute to Arbitration is a major decision because it will entail considerable costs in time and money which might never be recovered. It is because of this that the ICC is willing to act as a Conciliator if required to do so by the Parties in dispute, and to resolve the dispute before embarking on full Arbitration proceedings. Before Arbitration, both Parties will have obtained legal guidance about the laws applicable to the subject in the country concerned. These laws might well affect the selection of the nationality of the Arbitrators, the venue where such proceedings are to take place, and whether the award made by the Arbitrator can be challenged and overturned or if such an award is considered to be sacrosanct.

When both Parties reach the stage of entering into Arbitration, they have generally consulted eminent legal advisors who have no doubt assured them of victory. Nevertheless, any Party who is assured of a positive result in advance of the award should apply commercial expertise and judgement to evaluate the position if he should lose.

Before a dispute can be referred to Arbitration there are two important questions. First, does the Contract contain a Clause requiring disputes to be referred to Arbitration? This is a requirement of the ICC procedure and it needs to be confirmed that Clause 67 has not been deleted from the Contract Documents. Second, has the dispute been referred to the Engineer for a decision under Clause 67?

The procedure, once it is satisfied that Arbitration can take place, is as given in Chart 5. This traces the obligations and timing required to be observed, and their relationship with the requirements of the ICC Rules.

Despite his previous involvement as a decision-maker, the Engineer can, *Sub-Clause 67.3* under Sub-Clause 67.3, be called as a witness, or to give evidence to the Arbitrator(s). During the interim period when Arbitration is continuing, the Engineer is not precluded from conducting a dialogue with the Contractor or the Employer to seek a solution to the dispute, but he would have no authority to actually settle it.

The ICC Rules and Procedures[3] are available from the headquarters in Paris and probably offer the most economical and expeditious manner by which disputes can be resolved, along with a high degree of confidentiality, because the actions are not held in open court. The ICC can recommend on request experienced and available Arbitrators on most subjects through its National Committee.

There is a danger to a Contractor if he accepts an Arbitration Clause which already names the Arbitrator as an official of the same government or organisation as the Employer. This is not uncommon in certain countries where even a military governor can find a place in the system. A similar danger exists where the dictates of the ruling religion take precedence over the Laws of the Land, or where a particular calendar date can sometimes be of such significance and importance as to override the written word of the Contract.

To succeed in any Arbitration, it is essential for proper and detailed records, photographs, Site Diaries, correspondence, minutes, Drawings, maps, and such like to be made available to the Arbitrator. It is also helpful to remember that most Arbitrators for civil engineering Contracts have a good basic knowledge of the subject, but can often understand an argument more clearly if they are given charts and diagrams to support the written word.

Conclusions

The Digest forms an analysis of selected Clauses in the new fourth edition of the FIDIC *Conditions of contract for works of civil engineering construction* and explains the more important differences in this edition and the previous one. Relationships between the Parties to the Contract and others concerned with the Project have also been examined and various, hopefully useful, schedules and charts provided. Where the relationships and responsibilities are not observed, problems will arise and inevitably get out of proportion and become extremely difficult to resolve without one Party or other involving themselves in unnecessary additional expenditure. The very nature and complexity of undertaking works of civil engineering have proved that problems and claims will arise during the period of construction but provided these are dealt with as appropriate to the circumstances, as and when they occur, then a satisfactory conclusion can often be reached.

Users of the FIDIC *Conditions of contract* will recognise that frequency of use provides the best way of understanding the document. The Authors feel it incumbent upon themselves to remind such users that the most important feature of international construction is always to recognise the supremacy of the law of the country in which the Project is being constructed.

The new fourth edition has introduced many improvements of clarification and simplication to those Clauses which previously could have been misunderstood or confusing. It now makes the Employer more aware and involved than before as to what takes place during the construction period and affords considerable assistance to the Contractor on matters of interim payments, claims, and the Settlement of Disputes.

Comments from the readers of the Digest on any matters concerning the application of the Conditions as well as to any changes they consider desirable are welcome.

References

1. SAWYER J.G. and GILLOTT C.A. *The FIDIC Conditions. Digest of contractual relationships and responsibilities.* Thomas Telford, London, 1985, 2nd edn. [Covers 3rd edn FIDIC conditions.]
2. *FIDIC tendering procedure*, FIDIC, Lausanne.
3. INTERNATIONAL CHAMBER OF COMMERCE. *Rules of conciliation and arbitration.* ICC, 1988.
4. INTERNATIONAL CHAMBER OF COMMERCE. *Guide to arbitration.* ICC, 1983, publication 382.

Bibliography

FRICK-MEIJER S.-E. FIDIC conditions of contract for works of civil engineering construction, fourth edition. *Proc. Instn Civ. Engrs*, Part 1, 1988, **84**, Aug., 821–836 [seminar report].

MADGE P. *Civil engineering insurance and bonding.* Thomas Telford, London, 1987. (The book is not based on FIDIC conditions but is closely allied to them.]

FIDIC. *Guide to the use of FIDIC conditions of contract for works of civil engineering construction.* FIDIC, Lausanne, 1989, 4th edn.

FIDIC. Insurance of large civil engineering projects. FIDIC, Lausanne.

FIDIC. Construction, insurance and law. FIDIC, Lausanne, discussion paper.

FIDIC. *Conditions of contract for electrical and mechanical works (the Yellow Book).* FIDIC, Lausanne, 3rd edn.

FIDIC. *Guide to the use of FIDIC conditions for electrical and mechanical works.* FIDIC, Lausanne.

Appendix 1. Contractor's Site Diary — typical information recorded daily

The type of information that should be recorded by the Contractor in the Site Diary each day is set out in this appendix. The information recorded in the diaries of other Parties to the Contract will no doubt be very similar. The recorded data will be essential for establishing claims and counterclaims and facts concerning the Works.

Information required

External

1. Weather experienced during the past 24 hours, compared with forecast, plus a comment on any unusual unforeseen weather. *Sub-Clauses 20.4, 40.1, 44.1*

2. Note any news of external affairs likely to affect project e.g. strikes, hostilities, riots, acts of God such as earthquakes, hurricanes, and major accidents or catastrophies, unusual weather etc., whether in the country of the project or elsewhere. *Sub-Clauses 20.3, 44.1*

Works

1. Workforce — numbers of various categories of staff and labour on Site for main Contractor, Sub-Contractors and others. *Clause 35.1*

2. List of major plant and equipment on Site — company owned, hired and Sub-Contractor owned. *Clause 35.1*

3. Work being undertaken that day with special reference to major pours of concrete, heavy lifts, special installations etc.

4. Unexpected physical conditions encountered and action taken under Clause 12.2. *Sub-Clause 12.2*

5. Notes on any dangerous occurrences whether or not any damage or injury resulted, and on any accidents, fires, thefts, injuries, sickness or damage to property. *Sub-Clauses 20.1, 21.1, 24.1*

6. Note any correspondence with insurers and any claims made. *Sub-Clause 21.1*

7. Notes on receipt, and any delays experienced, of the following, including notes on news of impending delays

    (*a*) possession of Site and access thereto *Sub-Clauses 42.1, 42.2*

    (*b*) materials, temporary or permanent or consumable *Sub-Clause 44.1*
    (*c*) plant or equipment, temporary or permanent or on hire *Sub-Clause 44.1*

(*d*) Drawings

(*e* instructions

(*f*) information such as reports, test results etc.

Sub-Clauses 6.3, 6.4

8. Notes on consultations with Employer and Engineer.   *Section 3*

9. Notes on any labour unrest, strikes, rebellion, lockouts, with own and Sub-Contractors'/Suppliers' workforce and with Client's employees.   *Sub-Clause 44.1*

10. Discussions with union officials, shop stewards and other official or unofficial spokesmen for the labour force.

11. Discovery of fossils, coins, or articles of value or antiquity.   *Clause 27.1, Sub-Clause 44.1*

Engineer

1. Notes of any discussions with any Engineers' Representatives whether resident or visiting.

2. Reference to receipt or despatch of key letters or notices to Engineer or his Representatives.

3. Receipt of instructions from the Engineer or his Representatives whether verbal or in writing.   *Sub-Clauses 2.3, 2.5, 51.1, 51.2*

4. Despatch of written confirmation of verbal instructions.   *Sub-Clause 2.5*

5. Receipt of Drawings etc. from the Engineer.   *Sub-Clauses 6.2, 6.4*

6. Measurement

  (*a*) requests by Engineer to attend for measurement   *Clause 56.1*

  (*b*) failure of any side to attend.

7. Inspection

  (*a*) requests to attend inspection of Works, operations in workshops off site and of testing on and off site   *Clauses 37, 38, 39*

  (*b*) failure of any Party to attend.

8. Any information about invoices, certificates and payment.   *Sub-Clause 48.1, Clauses 60, 62.1*

9. Notes on all actions taken as required by Clause 53.   *Clause 53*

Contractor's Head Office

1. Discussions with any officials or visitors from Head Office.

2. Receipt and despatch of key letters etc.

3. Receipt of any important documents from Head Office by post, courier, telex/fax or by hand.

4. Information about invoices, certificates and payment.

Sub-Contractors and Suppliers — nominated or domestic   *Sub-Clauses 3.1, 59.4, 59.5*

1. Discussions with any Sub-Contractors, Suppliers or services representatives whether resident or not.

2. Receipt and despatch of key letters etc.

3. Issue of verbal or written instructions.

4. Confirmation of verbal instructions.

5. Measurement

   (*a*) requests to attend for measurement by either side   *Clause 56.1*
   (*b*) failure to attend.

6. Inspection

   (*a*) requests to attend for inspection of work   *Clauses 37, 38,*
   (*b*) failure to attend.   *39*

7. Information about invoices, certificates and payment.   *Sub-Clause 59.5*

Government officials, Employer's Representatives and bankers etc.
1. Notes of discussions with above.
2. Receipt and despatch of key letters etc.

Visitors to Site including press, TV and radio   *Sub-Clause 19.1*
1. Notes of all such visitors to Site recording names, purpose of visit and summary of activity undertaken.
2. Notes on discussions with visitors.

General
1. A special marking code should be adopted to identify any event or discussion which might give rise to a claim for Extension of Time and/or for additional costs.
2. If any shorthand form of notation is used then a legend explaining the meaning of any symbols must be displayed prominently in the front of the diary.
3. Wherever possible references should be added to identify letters, documents, Drawings, plant locations, Site locations etc.
4. In addition to any legends or reference codes used in the Diary it would be valuable to have a Site plan displayed as part of each volume of the Site Diary.
5. VERY IMPORTANT: Each page should be copied as completed and sent to Head Office in regular and frequent consignments and the original Diary must be kept in a fire-proof and vermin-proof safe under lock and key.
6. All Diaries must be kept for at least the period stated in the relevant Act of Limitation and possibly for at least fifteen years after completion of the Contract.

Appendix 2. FIDIC *Conditions of contract for works of civil engineering construction*, fourth edition, 1987

Part I General conditions with forms of tender and agreement, and Part II Conditions of particular application with guidelines for preparation of Part II clauses

CONTENTS

PART I: GENERAL CONDITIONS

Labour

Materials, Plant and Workmanship

Contractor's Equipment, Temporary Works and Materials

Measurement

Provisional Sums

Nominated Subcontractors

Certificates and Payment

Remedies

Special Risks

PART I – GENERAL CONDITIONS
Definitions and Interpretation

Definitions 1.1 In the Contract (as hereinafter defined) the following words and expressions shall have the meanings hereby assigned to them, except where the context otherwise requires:

(a) (i) "Employer" means the person named as such in Part II of these Conditions and the legal successors in title to such person, but not (except with the consent of the Contractor) any assignee of such person.

(ii) "Contractor" means the person whose tender has been accepted by the Employer and the legal successors in title to such person, but not (except with the consent of the Employer) any assignee of such person.

(iii) "Subcontractor" means any person named in the Contract as a Subcontractor for a part of the Works or any person to whom a part of the Works has been subcontracted with the consent of the Engineer and the legal successors in title to such person, but not any assignee of any such person.

(iv) "Engineer" means the person appointed by the Employer to act as Engineer for the purposes of the Contract and named as such in Part II of these Conditions.

(v) "Engineer's Representative" means a person appointed from time to time by the Engineer under Sub-Clause 2.2.

(b) (i) "Contract" means these Conditions (Parts I and II), the Specification, the Drawings, the Bill of Quantities, the Tender, the Letter of Acceptance, the Contract Agreement (if completed) and such further documents as may be expressly incorporated in the Letter of Acceptance or Contract Agreement (if completed).

(ii) "Specification" means the specification of the Works included in the Contract and any modification thereof or addition thereto made under Clause 51 or submitted by the Contractor and approved by the Engineer.

(iii) "Drawings" means all drawings, calculations and technical information of a like nature provided by the Engineer to the Contractor under the Contract and all drawings, calculations, samples, patterns, models, operation and maintenance manuals and other technical information of a like nature submitted by the Contractor and approved by the Engineer.

(iv) "Bill of Quantities" means the priced and completed bill of quantities forming part of the Tender.

(v) "Tender" means the Contractor's priced offer to the Employer for the execution and completion of the Works and the remedying of any defects therein in accordance with the provisions of the Contract, as accepted by the Letter of Acceptance.

(vi) "Letter of Acceptance" means the formal acceptance by the Employer of the Tender.

(vii) "Contract Agreement" means the contract agreement (if any) referred to in Sub-Clause 9.1.

(viii) "Appendix to Tender" means the appendix comprised in the form of Tender annexed to these Conditions.

(c) (i) "Commencement Date" means the date upon which the Contractor receives the notice to commence issued by the Engineer pursuant to Clause 41.

© FIDIC 1987

(ii) "Time for Completion" means the time for completing the execution of and passing the Tests on Completion of the Works or any Section or part thereof as stated in the Contract (or as extended under Clause 44) calculated from the Commencement Date.

(d)(i) "Tests on Completion" means the tests specified in the Contract or otherwise agreed by the Engineer and the Contractor which are to be made by the Contractor before the Works or any Section or part thereof are taken over by the Employer.

(ii) "Taking-Over Certificate" means a certificate issued pursuant to Clause 48.

(e)(i) "Contract Price" means the sum stated in the Letter of Acceptance as payable to the Contractor for the execution and completion of the Works and the remedying of any defects therein in accordance with the provisions of the Contract.

(ii) "Retention Money" means the aggregate of all monies retained by the Employer pursuant to Sub-Clause 60.2(a).

(f)(i) "Works" means the Permanent Works and the Temporary Works or either of them as appropriate.

(ii) "Permanent Works" means the permanent works to be executed (including Plant) in accordance with the Contract.

(iii) "Temporary Works" means all temporary works of every kind (other than Contractor's Equipment) required in or about the execution and completion of the Works and the remedying of any defects therein.

(iv) "Plant" means machinery, apparatus and the like intended to form or forming part of the Permanent Works.

(v) "Contractor's Equipment" means all appliances and things of whatsoever nature (other than Temporary Works) required for the execution and completion of the Works and the remedying of any defects therein, but does not include Plant, materials or other things intended to form or forming part of the Permanent Works.

(vi) "Section" means a part of the Works specifically identified in the Contract as a Section.

(vii) "Site" means the places provided by the Employer where the Works are to be executed and any other places as may be specifically designated in the Contract as forming part of the Site.

(g)(i) "cost" means all expenditure properly incurred or to be incurred, whether on or off the Site, including overhead and other charges properly allocable thereto but does not include any allowance for profit.

(ii) "day" means calendar day.

(iii) "foreign currency" means a currency of a country other than that in which the Works are to be located.

(iv) "writing" means any hand-written, type-written, or printed communication, including telex, cable and facsimile transmission.

Headings and Marginal Notes   1.2   The headings and marginal notes in these Conditions shall not be deemed part thereof or be taken into consideration in the interpretation or construction thereof or of the Contract.

Interpretation   1.3   Words importing persons or parties shall include firms and corporations and any organisation having legal capacity.

Singular and Plural   1.4   Words importing the singular only also include the plural and vice versa where the context requires.

| | **1.5** | Wherever in the Contract provision is made for the giving or issue of any notice, consent, approval, certificate or determination by any person, unless otherwise specified such notice, consent, approval, certificate or determination shall be in writing and the words "notify", "certify" or "determine" shall be construed accordingly. Any such consent, approval, certificate or determination shall not unreasonably be withheld or delayed. |

Notices, Consents, Approvals, Certificates and Determinations

Engineer and Engineer's Representative

Engineer's Duties and Authority **2.1** (a) The Engineer shall carry out the duties specified in the Contract.

(b) The Engineer may exercise the authority specified in or necessarily to be implied from the Contract, provided, however, that if the Engineer is required, under the terms of his appointment by the Employer, to obtain the specific approval of the Employer before exercising any such authority, particulars of such requirements shall be set out in Part II of these Conditions. Provided further that any requisite approval shall be deemed to have been given by the Employer for any such authority exercised by the Engineer.

(c) Except as expressly stated in the Contract, the Engineer shall have no authority to relieve the Contractor of any of his obligations under the Contract.

Engineer's Representative **2.2** The Engineer's Representative shall be appointed by and be responsible to the Engineer and shall carry out such duties and exercise such authority as may be delegated to him by the Engineer under Sub-Clause 2.3.

Engineer's Authority to Delegate **2.3** The Engineer may from time to time delegate to the Engineer's Representative any of the duties and authorities vested in the Engineer and he may at any time revoke such delegation. Any such delegation or revocation shall be in writing and shall not take effect until a copy thereof has been delivered to the Employer and the Contractor.

Any communication given by the Engineer's Representative to the Contractor in accordance with such delegation shall have the same effect as though it had been given by the Engineer. Provided that:

(a) any failure of the Engineer's Representative to disapprove any work, materials or Plant shall not prejudice the authority of the Engineer to disapprove such work, materials or Plant and to give instructions for the rectification thereof;

(b) if the Contractor questions any communication of the Engineer's Representative he may refer the matter to the Engineer who shall confirm, reverse or vary the contents of such communication.

Appointment of Assistants **2.4** The Engineer or the Engineer's Representative may appoint any number of persons to assist the Engineer's Representative in the carrying out of his duties under Sub-Clause 2.2. He shall notify to the Contractor the names, duties and scope of authority of such persons. Such assistants shall have no authority to issue any instructions to the Contractor save in so far as such instructions may be necessary to enable them to carry out their duties and to secure their acceptance of materials, Plant or workmanship as being in accordance with the Contract, and any instructions given by any of them for those purposes shall be deemed to have been given by the Engineer's Representative.

Instructions in Writing **2.5** Instructions given by the Engineer shall be in writing, provided that if for any reason the Engineer considers it necessary to give any such instruction orally, the Contractor shall comply with such instruction. Confirmation in writing of such oral instruction given by the Engineer, whether before or after the carrying out of the instruction, shall be deemed to be an instruction within the meaning of this Sub-Clause. Provided further that if the Contractor, within 7 days, confirms in writing to the Engineer any oral instruction of the Engineer and such confirmation is not contradicted in writing within 7 days by the Engineer, it shall be deemed to be an instruction of the Engineer.

The provisions of this Sub-Clause shall equally apply to instructions given by the Engineer's Representative and any assistants of the Engineer or the Engineer's Representative appointed pursuant to Sub-Clause 2.4.

Engineer to Act Impartially 2.6 Wherever, under the Contract, the Engineer is required to exercise his discretion by:

(a) giving his decision, opinion or consent, or

(b) expressing his satisfaction or approval, or

(c) determining value, or

(d) otherwise taking action which may affect the rights and obligations of the Employer or the Contractor

he shall exercise such discretion impartially within the terms of the Contract and having regard to all the circumstances. Any such decision, opinion, consent, expression of satisfaction, or approval, determination of value or action may be opened up, reviewed or revised as provided in Clause 67.

Assignment and Subcontracting

Assignment of Contract 3.1 The Contractor shall not, without the prior consent of the Employer (which consent, notwithstanding the provisions of Sub-Clause 1.5, shall be at the sole discretion of the Employer), assign the Contract or any part thereof, or any benefit or interest therein or thereunder, otherwise than by:

(a) a charge in favour of the Contractor's bankers of any monies due or to become due under the Contract, or

(b) assignment to the Contractor's insurers (in cases where the insurers have discharged the Contractor's loss or liability) of the Contractor's right to obtain relief against any other party liable.

Subcontracting 4.1 The Contractor shall not subcontract the whole of the Works. Except where otherwise provided by the Contract, the Contractor shall not subcontract any part of the Works without the prior consent of the Engineer. Any such consent shall not relieve the Contractor from any liability or obligation under the Contract and he shall be responsible for the acts, defaults and neglects of any Subcontractor, his agents, servants or workmen as fully as if they were the acts, defaults or neglects of the Contractor, his agents, servants or workmen.

Provided that the Contractor shall not be required to obtain such consent for:

(a) the provision of labour, or

(b) the purchase of materials which are in accordance with the standards specified in the Contract, or

(c) the subcontracting of any part of the Works for which the Subcontractor is named in the Contract.

Assignment of Subcontractors' Obligations 4.2 In the event of a Subcontractor having undertaken towards the Contractor in respect of the work executed, or the goods, materials, Plant or services supplied by such Subcontractor, any continuing obligation extending for a period exceeding that of the Defects Liability Period under the Contract, the Contractor shall at any time, after the expiration of such Period, assign to the Employer, at the Employer's request and cost, the benefit of such obligation for the unexpired duration thereof.

Contract Documents

Language/s and Law 5.1 There is stated in Part II of these Conditions:

(a) the language or languages in which the Contract documents shall be drawn up, and

© FIDIC 1987

(b) the country or state the law of which shall apply to the Contract and according to which the Contract shall be construed.

If the said documents are written in more than one language, the language according to which the Contract shall be construed and interpreted is also stated in Part II of these Conditions, being therein designated the "Ruling Language".

Priority of Contract Documents

5.2 The several documents forming the Contract are to be taken as mutually explanatory of one another, but in case of ambiguities or discrepancies the same shall be explained and adjusted by the Engineer who shall thereupon issue to the Contractor instructions thereon and in such event, unless otherwise provided in the Contract, the priority of the documents forming the Contract shall be as follows:

(1) The Contract Agreement (if completed);

(2) The Letter of Acceptance;

(3) The Tender;

(4) Part II of these Conditions;

(5) Part I of these Conditions; and

(6) Any other document forming part of the Contract.

Custody and Supply of Drawings and Documents

6.1 The Drawings shall remain in the sole custody of the Engineer, but two copies thereof shall be provided to the Contractor free of charge. The Contractor shall make at his own cost any further copies required by him. Unless it is strictly necessary for the purposes of the Contract, the Drawings, Specification and other documents provided by the Employer or the Engineer shall not, without the consent of the Engineer, be used or communicated to a third party by the Contractor. Upon issue of the Defects Liability Certificate, the Contractor shall return to the Engineer all Drawings, Specification and other documents provided under the Contract.

The Contractor shall supply to the Engineer four copies of all Drawings, Specification and other documents submitted by the Contractor and approved by the Engineer in accordance with Clause 7, together with a reproducible copy of any material which cannot be reproduced to an equal standard by photocopying. In addition the Contractor shall supply such further copies of such Drawings, Specification and other documents as the Engineer may request in writing for the use of the Employer, who shall pay the cost thereof.

One Copy of Drawings to be Kept on Site

6.2 One copy of the Drawings, provided to or supplied by the Contractor as aforesaid, shall be kept by the Contractor on the Site and the same shall at all reasonable times be available for inspection and use by the Engineer and by any other person authorised by the Engineer in writing.

Disruption of Progress

6.3 The Contractor shall give notice to the Engineer, with a copy to the Employer, whenever planning or execution of the Works is likely to be delayed or disrupted unless any further drawing or instruction is issued by the Engineer within a reasonable time. The notice shall include details of the drawing or instruction required and of why and by when it is required and of any delay or disruption likely to be suffered if it is late.

Delays and Cost of Delay of Drawings

6.4 If, by reason of any failure or inability of the Engineer to issue, within a time reasonable in all the circumstances, any drawing or instruction for which notice has been given by the Contractor in accordance with Sub-Clause 6.3, the Contractor suffers delay and/or incurs costs then the Engineer shall, after due consultation with the Employer and the Contractor, determine:

(a) any extension of time to which the Contractor is entitled under Clause 44, and

(b) the amount of such costs, which shall be added to the Contract Price,

and shall notify the Contractor accordingly, with a copy to the Employer.

© FIDIC 1987

Failure by Contractor to Submit Drawings 6.5 If the failure or inability of the Engineer to issue any drawings or instructions is caused in whole or in part by the failure of the Contractor to submit Drawings, Specification or other documents which he is required to submit under the Contract, the Engineer shall take such failure by the Contractor into account when making his determination pursuant to Sub-Clause 6.4.

Supplementary Drawings and Instructions 7.1 The Engineer shall have authority to issue to the Contractor, from time to time, such supplementary Drawings and instructions as shall be necessary for the purpose of the proper and adequate execution and completion of the Works and the remedying of any defects therein. The Contractor shall carry out and be bound by the same.

Permanent Works Designed by Contractor 7.2 Where the Contract expressly provides that part of the Permanent Works shall be designed by the Contractor, he shall submit to the Engineer, for approval:

(a) such drawings, specifications, calculations and other information as shall be necessary to satisfy the Engineer as to the suitability and adequacy of that design, and

(b) operation and maintenance manuals together with drawings of the Permanent Works as completed, in sufficient detail to enable the Employer to operate, maintain, dismantle, reassemble and adjust the Permanent Works incorporating that design. The Works shall not be considered to be completed for the purposes of taking over in accordance with Clause 48 until such operation and maintenance manuals, together with drawings on completion, have been submitted to and approved by the Engineer.

Responsibility Unaffected by Approval 7.3 Approval by the Engineer, in accordance with Sub-Clause 7.2, shall not relieve the Contractor of any of his responsibilities under the Contract.

General Obligations

Contractor's General Responsibilities 8.1 The Contractor shall, with due care and diligence, design (to the extent provided for by the Contract), execute and complete the Works and remedy any defects therein in accordance with the provisions of the Contract. The Contractor shall provide all superintendence, labour, materials, Plant, Contractor's Equipment and all other things, whether of a temporary or permanent nature, required in and for such design, execution, completion and remedying of any defects, so far as the necessity for providing the same is specified in or is reasonably to be inferred from the Contract.

Site Operations and Methods of Construction 8.2 The Contractor shall take full responsibility for the adequacy, stability and safety of all Site operations and methods of construction. Provided that the Contractor shall not be responsible (except as stated hereunder or as may be otherwise agreed) for the design or specification of Permanent Works, or. for the design or specification of any Temporary Works not prepared by the Contractor. Where the Contract expressly provides that part of the Permanent Works shall be designed by the Contractor, he shall be fully responsible for that part of such Works, notwithstanding any approval by the Engineer.

Contract Agreement 9.1 The Contractor shall, if called upon so to do, enter into and execute the Contract Agreement, to be prepared and completed at the cost of the Employer, in the form annexed to these Conditions with such modification as may be necessary.

Performance Security 10.1 If the Contract requires the Contractor to obtain security for his proper performance of the Contract he shall obtain and provide to the Employer such security within 28 days after the receipt of the Letter of Acceptance, in the sum stated in the Appendix to Tender. When providing such security to the Employer, the Contractor shall notify the Engineer of so doing. Such security shall be in such form as may be agreed between the Employer and the Contractor. The institution providing such security shall be subject to the approval of the Employer. The cost of complying with the requirements of this Clause shall be borne by the Contractor, unless the Contract otherwise provides.

Period of Validity of Performance Security	**10.2**	The performance security shall be valid until the Contractor has executed and completed the Works and remedied any defects therein in accordance with the Contract. No claim shall be made against such security after the issue of the Defects Liability Certificate in accordance with Sub-Clause 62.1 and such security shall be returned to the Contractor within 14 days of the issue of the said Defects Liability Certificate.
Claims under Performance Security	**10.3**	Prior to making a claim under the performance security the Employer shall, in every case, notify the Contractor stating the nature of the default in respect of which the claim is to be made.
Inspection of Site	**11.1**	The Employer shall have made available to the Contractor, before the submission by the Contractor of the Tender, such data on hydrological and sub-surface conditions as have been obtained by or on behalf of the Employer from investigations undertaken relevant to the Works but the Contractor shall be responsible for his own interpretation thereof.

The Contractor shall be deemed to have inspected and examined the Site and its surroundings and information available in connection therewith and to have satisfied himself (so far as is practicable, having regard to considerations of cost and time) before submitting his Tender, as to:

(a) the form and nature thereof, including the sub-surface conditions,

(b) the hydrological and climatic conditions,

(c) the extent and nature of work and materials necessary for the execution and completion of the Works and the remedying of any defects therein, and

(d) the means of access to the Site and the accommodation he may require

and, in general, shall be deemed to have obtained all necessary information, subject as above mentioned, as to risks, contingencies and all other circumstances which may influence or affect his Tender.

The Contractor shall be deemed to have based his Tender on the data made available by the Employer and on his own inspection and examination, all as aforementioned.

Sufficiency of Tender	**12.1**	The Contractor shall be deemed to have satisfied himself as to the correctness and sufficiency of the Tender and of the rates and prices stated in the Bill of Quantities, all of which shall, except insofar as it is otherwise provided in the Contract, cover all his obligations under the Contract (including those in respect of the supply of goods, materials, Plant or services or of contingencies for which there is a Provisional Sum) and all matters and things necessary for the proper execution and completion of the Works and the remedying of any defects therein.
Adverse Physical Obstructions or Conditions	**12.2**	If, however, during the execution of the Works the Contractor encounters physical obstructions or physical conditions, other than climatic conditions on the Site, which obstructions or conditions were, in his opinion, not foreseeable by an experienced contractor, the Contractor shall forthwith give notice thereof to the Engineer, with a copy to the Employer. On receipt of such notice, the Engineer shall, if in his opinion such obstructions or conditions could not have been reasonably foreseen by an experienced contractor, after due consultation with the Employer and the Contractor, determine:

(a) any extension of time to which the Contractor is entitled under Clause 44, and

(b) the amount of any costs which may have been incurred by the Contractor by reason of such obstructions or conditions having been encountered, which shall be added to the Contract Price,

and shall notify the Contractor accordingly, with a copy to the Employer. Such determination shall take account of any instruction which the Engineer may issue to the Contractor in connection therewith, and any proper and reasonable measures acceptable to the Engineer which the Contractor may take in the absence of specific instructions from the Engineer.

Work to be in Accordance with Contract	**13.1**	Unless it is legally or physically impossible, the Contractor shall execute and complete the Works and remedy any defects therein in strict accordance with the Contract to the satisfaction of the Engineer. The Contractor shall comply with and adhere strictly to the Engineer's instructions on any matter, whether mentioned in the Contract or not, touching or concerning the Works. The Contractor shall take instructions only from the Engineer or, subject to the provisions of Clause 2, from the Engineer's Representative.
Programme to be Submitted	**14.1**	The Contractor shall, within the time stated in Part II of these Conditions after the date of the Letter of Acceptance, submit to the Engineer for his consent a programme, in such form and detail as the Engineer shall reasonably prescribe, for the execution of the Works. The Contractor shall, whenever required by the Engineer, also provide in writing for his information a general description of the arrangements and methods which the Contractor proposes to adopt for the execution of the Works.
Revised Programme	**14.2**	If at any time it should appear to the Engineer that the actual progress of the Works does not conform to the programme to which consent has been given under Sub-Clause 14.1, the Contractor shall produce, at the request of the Engineer, a revised programme showing the modifications to such programme necessary to ensure completion of the Works within the Time for Completion.
Cash Flow Estimate to be Submitted	**14.3**	The Contractor shall, within the time stated in Part II of these Conditions after the date of the Letter of Acceptance, provide to the Engineer for his information a detailed cash flow estimate, in quarterly periods, of all payments to which the Contractor will be entitled under the Contract and the Contractor shall subsequently supply revised cash flow estimates at quarterly intervals, if required to do so by the Engineer.
Contractor not Relieved of Duties or Responsibilities	**14.4**	The submission to and consent by the Engineer of such programmes or the provision of such general descriptions or cash flow estimates shall not relieve the Contractor of any of his duties or responsibilities under the Contract.
Contractor's Superintendence	**15.1**	The Contractor shall provide all necessary superintendence during the execution of the Works and as long thereafter as the Engineer may consider necessary for the proper fulfilling of the Contractor's obligations under the Contract. The Contractor, or a competent and authorised representative approved of by the Engineer, which approval may at any time be withdrawn, shall give his whole time to the superintendence of the Works. Such authorised representative shall receive, on behalf of the Contractor, instructions from the Engineer or, subject to the provisions of Clause 2, the Engineer's Representative.

If approval of the representative is withdrawn by the Engineer, the Contractor shall, as soon as is practicable, having regard to the requirement of replacing him as hereinafter mentioned, after receiving notice of such withdrawal, remove the representative from the Works and shall not thereafter employ him again on the Works in any capacity and shall replace him by another representative approved by the Engineer.

Contractor's Employees	**16.1**	The Contractor shall provide on the Site in connection with the execution and completion of the Works and the remedying of any defects therein

(a) only such technical assistants as are skilled and experienced in their respective callings and such foremen and leading hands as are competent to give proper superintendence of the Works, and

(b) such skilled, semi-skilled and unskilled labour as is necessary for the proper and timely fulfilling of the Contractor's obligations under the Contract.

Engineer at Liberty to Object	**16.2**	The Engineer shall be at liberty to object to and require the Contractor to remove forthwith from the Works any person provided by the Contractor who, in the opinion of the Engineer, misconducts himself, or is incompetent or negligent in the proper performance of his duties, or whose presence on Site is otherwise considered by the Engineer to be undesirable, and such person shall not be again allowed upon the Works without the consent of the Engineer. Any person so removed from the Works shall be replaced as soon as possible.

Setting-out	**17.1**	The Contractor shall be responsible for:

(a) the accurate setting-out of the Works in relation to original points, lines and levels of reference given by the Engineer in writing,

(b) the correctness, subject as above mentioned, of the position, levels, dimensions and alignment of all parts of the Works, and

(c) the provision of all necessary instruments, appliances and labour in connection with the foregoing responsibilities.

If, at any time during the execution of the Works, any error appears in the position, levels, dimensions or alignment of any part of the Works, the Contractor, on being required so to do by the Engineer, shall, at his own cost, rectify such error to the satisfaction of the Engineer, unless such error is based on incorrect data supplied in writing by the Engineer, in which case the Engineer shall determine an addition to the Contract Price in accordance with Clause 52 and shall notify the Contractor accordingly, with a copy to the Employer.

The checking of any setting-out or of any line or level by the Engineer shall not in any way relieve the Contractor of his responsibility for the accuracy thereof and the Contractor shall carefully protect and preserve all bench-marks, sight-rails, pegs and other things used in setting-out the Works.

Boreholes and Exploratory Excavation	**18.1**	If, at any time during the execution of the Works, the Engineer requires the Contractor to make boreholes or to carry out exploratory excavation, such requirement shall be the subject of an instruction in accordance with Clause 51, unless an item or a Provisional Sum in respect of such work is included in the Bill of Quantities.

Safety, Security and Protection of the Environment	**19.1**	The Contractor shall, throughout the execution and completion of the Works and the remedying of any defects therein:

(a) have full regard for the safety of all persons entitled to be upon the Site and keep the Site (so far as the same is under his control) and the Works (so far as the same are not completed or occupied by the Employer) in an orderly state appropriate to the avoidance of danger to such persons, and

(b) provide and maintain at his own cost all lights, guards, fencing, warning signs and watching, when and where necessary or required by the Engineer or by any duly constituted authority, for the protection of the Works or for the safety and convenience of the public or others, and

(c) take all reasonable steps to protect the environment on and off the Site and to avoid damage or nuisance to persons or to property of the public or others resulting from pollution, noise or other causes arising as a consequence of his methods of operation.

Employer's Responsibilities	**19.2**	If under Clause 31 the Employer shall carry out work on the Site with his own workmen he shall, in respect of such work:

(a) have full regard to the safety of all persons entitled to be upon the Site, and

(b) keep the Site in an orderly state appropriate to the avoidance of danger to such persons.

If under Clause 31 the Employer shall employ other contractors on the Site he shall require them to have the same regard for safety and avoidance of danger.

Care of Works	**20.1**	The Contractor shall take full responsibility for the care of the Works and materials and Plant for incorporation therein from the Commencement Date until the date of issue of the Taking-Over Certificate for the whole of the Works, when the responsibility for the said care shall pass to the Employer. Provided that:

(a) if the Engineer issues a Taking-Over Certificate for any Section or part of the Permanent Works the Contractor shall cease to be liable for the care of that Section or part from the date of issue of the Taking-Over Certificate, when the responsibility for the care of that Section or part shall pass to the Employer, and

(b) the Contractor shall take full responsibility for the care of any outstanding Works and materials and Plant for incorporation therein which he undertakes to finish during the Defects Liability Period until such outstanding Works have been completed pursuant to Clause 49.

Responsibility to Rectify Loss or Damage   **20.2**   If any loss or damage happens to the Works, or any part thereof, or materials or Plant for incorporation therein, during the period for which the Contractor is responsible for the care thereof, from any cause whatsoever, other than the risks defined in Sub-Clause 20.4, the Contractor shall, at his own cost, rectify such loss or damage so that the Permanent Works conform in every respect with the provisions of the Contract to the satisfaction of the Engineer. The Contractor shall also be liable for any loss or damage to the Works occasioned by him in the course of any operations carried out by him for the purpose of complying with his obligations under Clauses 49 and 50.

Loss or Damage Due to Employer's Risks   **20.3**   In the event of any such loss or damage happening from any of the risks defined in Sub-Clause 20.4, or in combination with other risks, the Contractor shall, if and to the extent required by the Engineer, rectify the loss or damage and the Engineer shall determine an addition to the Contract Price in accordance with Clause 52 and shall notify the Contractor accordingly, with a copy to the Employer. In the case of a combination of risks causing loss or damage any such determination shall take into account the proportional responsibility of the Contractor and the Employer.

Employer's Risks   **20.4**   The Employer's risks are:

(a) war, hostilities (whether war be declared or not), invasion, act of foreign enemies,

(b) rebellion, revolution, insurrection, or military or usurped power, or civil war,

(c) ionising radiations, or contamination by radio-activity from any nuclear fuel, or from any nuclear waste from the combustion of nuclear fuel, radio-active toxic explosive, or other hazardous properties of any explosive nuclear assembly or nuclear component thereof,

(d) pressure waves caused by aircraft or other aerial devices travelling at sonic or supersonic speeds,

(e) riot, commotion or disorder, unless solely restricted to employees of the Contractor or of his Subcontractors and arising from the conduct of the Works,

(f) loss or damage due to the use or occupation by the Employer of any Section or part of the Permanent Works, except as may be provided for in the Contract,

(g) loss or damage to the extent that it is due to the design of the Works, other than any part of the design provided by the Contractor or for which the Contractor is responsible,

(h) any operation of the forces of nature against which an experienced contractor could not reasonably have been expected to take precautions.

Insurance of Works and Contractor's Equipment   **21.1**   The Contractor shall, without limiting his or the Employer's obligations and responsibilities under Clause 20, insure:

(a) the Works, together with materials and Plant for incorporation therein, to the full replacement cost

(b) an additional sum of 15 per cent of such replacement cost, or as may be specified in Part II of these Conditions, to cover any additional costs of and incidental to the rectification of loss or damage including professional fees and the cost of demolishing and removing any part of the Works and of removing debris of whatsoever nature

(c) the Contractor's Equipment and other things brought onto the Site by the Contractor, for a sum sufficient to provide for their replacement at the Site.

© FIDIC 1987

Scope of Cover 21.2 The insurance in paragraphs (a) and (b) of Sub-Clause 21.1 shall be in the joint names of the Contractor and the Employer and shall cover:

(a) the Employer and the Contractor against all loss or damage from whatsoever cause arising, other than as provided in Sub-Clause 21.4, from the start of work at the Site until the date of issue of the relevant Taking-Over Certificate in respect of the Works or any Section or part thereof as the case may be, and

(b) the Contractor for his liability:

(i) during the Defects Liability Period for loss or damage arising from a cause occurring prior to the commencement of the Defects Liability Period, and

(ii) for loss or damage occasioned by the Contractor in the course of any operations carried out by him for the purpose of complying with his obligations under Clauses 49 and 50. .

Responsibility for 21.3 Any amounts not insured or not recovered from the insurers shall be borne by the
Amounts not Employer or the Contractor in accordance with their responsibilities under
Recovered Clause 20.

Exclusions 21.4 There shall be no obligation for the insurances in Sub-Clause 21.1 to include loss or damage caused by

(a) war, hostilities (where war be declared or not), invasion, act of foreign enemies,

(b) rebellion, revolution, insurrection, or military or usurped power, or civil war,

(c) ionising radiations, or contamination by radio-activity from any nuclear fuel, or from any nuclear waste from the combustion of nuclear fuel, radio-active toxic explosive, or other hazardous properties of any explosive nuclear assembly or nuclear component thereof,

(d) pressure waves caused by aircraft or other aerial devices travelling at sonic or supersonic speeds.

Damage to 22.1 The Contractor shall, except if and so far as the Contract provides otherwise,
Persons and indemnify the Employer against all losses and claims in respect of:
Property
(a) death of or injury to any person, or

(b) loss of or damage to any property (other than the Works), which may arise out of or in consequence of the execution and completion of the Works and the remedying of any defects therein, and against all claims, proceedings, damages, costs, charges and expenses whatsoever in respect thereof or in relation thereto, subject to the exceptions defined in Sub-Clause 22.2.

Exceptions 22.2 The "exceptions" referred to in Sub-Clause 22.1 are:

(a) the permanent use or occupation of land by the Works, or any part thereof,

(b) the right of the Employer to execute the Works, or any part thereof, on, over, under, in or through any land,

(c) damage to property which is the unavoidable result of the execution and completion of the Works, or the remedying of any defects therein, in accordance with the Contract,

(d) death of or injury to persons or loss of or damage to property resulting from any act or neglect of the Employer, his agents, servants or other contractors, not being employed by the Contractor, or in respect of any claims, proceedings, damages, costs, charges and expenses in respect thereof or in relation thereto or, where the injury or damage was contributed to by the Contractor, his servants or agents, such part of the said injury or damage as may be just and equitable having regard to the extent of the responsibility of the Employer, his servants or agents or other contractors for the injury or damage.

Indemnity by Employer	22.3	The Employer shall indemnify the Contractor against all claims, proceedings, damages, costs, charges and expenses in respect of the matters referred to in the exceptions defined in Sub-Clause 22.2.
Third Party Insurance (including Employer's Property)	23.1	The Contractor shall, without limiting his or the Employer's obligations and responsibilities under Clause 22, insure, in the joint names of the Contractor and the Employer, against liabilities for death of or injury to any person (other than as provided in Clause 24) or loss of or damage to any property (other than the Works) arising out of the performance of the Contract, other than the exceptions defined in paragraphs (a), (b) and (c) of Sub-Clause 22.2.
Minimum Amount of Insurance	23.2	Such insurance shall be for at least the amount stated in the Appendix to Tender.
Cross Liabilities	23.3	The insurance policy shall include a cross liability clause such that the insurance shall apply to the Contractor and to the Employer as separate insureds.
Accident or Injury to Workmen	24.1	The Employer shall not be liable for or in respect of any damages or compensation payable to any workman or other person in the employment of the Contractor or any Subcontractor, other than death or injury resulting from any act or default of the Employer, his agents or servants. The Contractor shall indemnify and keep indemnified the Employer against all such damages and compensation, other than those for which the Employer is liable as aforesaid, and against all claims, proceedings, damages, costs, charges, and expenses whatsoever in respect thereof or in relation thereto.
Insurance Against Accident to Workmen	24.2	The Contractor shall insure against such liability and shall continue such insurance during the whole of the time that any persons are employed by him on the Works. Provided that, in respect of any persons employed by any Subcontractor, the Contractor's obligations to insure as aforesaid under this Sub-Clause shall be satisfied if the Subcontractor shall have insured against the liability in respect of such persons in such manner that the Employer is indemnified under the policy, but the Contractor shall require such Subcontractor to produce to the Employer, when required, such policy of insurance and the receipt for the payment of the current premium.
Evidence and Terms of Insurances	25.1	The Contractor shall provide evidence to the Employer prior to the start of work at the Site that the insurances required under the Contract have been effected and shall, within 84 days of the Commencement Date, provide the insurance policies to the Employer. When providing such evidence and such policies to the Employer, the Contractor shall notify the Engineer of so doing. Such insurance policies shall be consistent with the general terms agreed prior to the issue of the Letter of Acceptance. The Contractor shall effect all insurances for which he is responsible with insurers and in terms approved by the Employer.
Adequacy of Insurances	25.2	The Contractor shall notify the insurers of changes in the nature, extent or programme for the execution of the Works and ensure the adequacy of the insurances at all times in accordance with the terms of the Contract and shall, when required, produce to the Employer the insurance policies in force and the receipts for payment of the current premiums.
Remedy on Contractor's Failure to Insure	25.3	If the Contractor fails to effect and keep in force any of the insurances required under the Contract, or fails to provide the policies to the Employer within the period required by Sub-Clause 25.1, then and in any such case the Employer may effect and keep in force any such insurances and pay any premium as may be necessary for that purpose and from time to time deduct the amount so paid from any monies due or to become due to the Contractor, or recover the same as a debt due from the Contractor.
Compliance with Policy Conditions	25.4	In the event that the Contractor or the Employer fails to comply with conditions imposed by the insurance policies effected pursuant to the Contract, each shall indemnify the other against all losses and claims arising from such failure.

Compliance with Statutes, Regulations	26.1	The Contractor shall conform in all respects, including by the giving of all notices and the paying of all fees, with the provisions of:

(a) any National or State Statute, Ordinance, or other Law, or any regulation, or bye-law of any local or other duly constituted authority in relation to the execution and completion of the Works and the remedying of any defects therein, and

(b) the rules and regulations of all public bodies and companies whose property or rights are affected or may be affected in any way by the Works,

and the Contractor shall keep the Employer indemnified against all penalties and liability of every kind for breach of any such provisions. Provided always that the Employer shall be responsible for obtaining any planning, zoning or other similar permission required for the Works to proceed and shall indemnify the Contractor in accordance with Sub-Clause 22.3.

Fossils 27.1 All fossils, coins, articles of value or antiquity and structures and other remains or things of geological or archaeological interest discovered on the Site shall, as between the Employer and the Contractor, be deemed to be the absolute property of the Employer. The Contractor shall take reasonable precautions to prevent his workmen or any other persons from removing or damaging any such article or thing and shall, immediately upon discovery thereof and before removal, acquaint the Engineer of such discovery and carry out the Engineer's instructions for dealing with the same. If, by reason of such instructions, the Contractor suffers delay and/or incurs costs then the Engineer shall, after due consultation with the Employer and the Contractor, determine:

(a) any extension of time to which the Contractor is entitled under Clause 44, and

(b) the amount of such costs, which shall be added to the Contract Price,

and shall notify the Contractor accordingly, with a copy to the Employer.

Patent Rights 28.1 The Contractor shall save harmless and indemnify the Employer from and against all claims and proceedings for or on account of infringement of any patent rights, design trademark or name or other protected rights in respect of any Contractor's Equipment, materials or Plant used for or in connection with or for incorporation in the Works and from and against all damages, costs, charges and expenses whatsoever in respect thereof or in relation thereto, except where such infringement results from compliance with the design or Specification provided by the Engineer.

Royalties 28.2 Except where otherwise stated, the Contractor shall pay all tonnage and other royalties, rent and other payments or compensation, if any, for getting stone, sand, gravel, clay or other materials required for the Works.

Interference with Traffic and Adjoining Properties 29.1 All operations necessary for the execution and completion of the Works and the remedying of any defects therein shall, so far as compliance with the requirements of the Contract permits, be carried on so as not to interfere unnecessarily or improperly with:

(a) the convenience of the public, or

(b) the access to, use and occupation of public or private roads and footpaths to or of properties whether in the possession of the Employer or of any other person.

The Contractor shall save harmless and indemnify the Employer in respect of all claims, proceedings, damages, costs, charges and expenses whatsoever arising out of, or in relation to, any such matters insofar as the Contractor is responsible therefor.

Avoidance of Damage to Roads 30.1 The Contractor shall use every reasonable means to prevent any of the roads or bridges communicating with or on the routes to the Site from being damaged or injured by any traffic of the Contractor or any of his Subcontractors and, in particular , shall select routes, choose and use vehicles and restrict and distribute loads so that any such extraordinary traffic as will inevitably arise from the moving of materials, Plant, Contractor's Equipment or Temporary Works from and to the Site shall be limited, as far as reasonably possible, and so that no unnecessary damage or injury may be occasioned to such roads and bridges.

Transport of Contractor's Equipment or Temporary Works 30.2 Save insofar as the Contract otherwise provides, the Contractor shall be responsible for and shall pay the cost of strengthening any bridges or altering or improving any road communicating with or on the routes to the Site to facilitate the movement of Contractor's Equipment or Temporary Works and the Contractor shall indemnify and keep indemnified the Employer against all claims for damage to any such road or bridge caused by such movement, including such claims as may be made directly against the Employer, and shall negotiate and pay all claims arising solely out of such damage.

Transport of Materials or Plant 30.3 If, notwithstanding Sub-Clause 30.1, any damage occurs to any bridge or road communicating with or on the routes to the Site arising from the transport of materials or Plant, the Contractor shall notify the Engineer with a copy to the Employer, as soon as he becomes aware of such damage or as soon as he receives any claim from the authority entitled to make such claim. Where under any law or regulation the haulier of such materials or Plant is required to indemnify the road authority against damage the Employer shall not be liable for any costs, charges or expenses in respect thereof or in relation thereto. In other cases the Employer shall negotiate the settlement of and pay all sums due in respect of such claim and shall indemnify the Contractor in respect thereof and in respect of all claims, proceedings, damages, costs, charges and expenses in relation thereto. Provided that if and so far as any such claim or part thereof is, in the opinion of the Engineer, due to any failure on the part of the Contractor to observe and perform his obligations under Sub-Clause 30.1, then the amount, determined by the Engineer, after due consultation with the Employer and the Contractor, to be due to such failure shall be recoverable from the Contractor by the Employer and may be deducted by the Employer from any monies due or to become due to the Contractor and the Engineer shall notify the Contractor accordingly, with a copy to the Employer. Provided also that the Employer shall notify the Contractor whenever a settlement is to be negotiated and, where any amount may be due from the Contractor, the Employer shall consult with the Contractor before such settlement is agreed.

Waterborne Traffic 30.4 Where the nature of the Works is such as to require the use by the Contractor of waterborne transport the foregoing provisions of this Clause shall be construed as though "road" included a lock, dock, sea wall or other structure related to a waterway and "vehicle" included craft, and shall have effect accordingly.

Opportunities for Other Contractors 31.1 The Contractor shall, in accordance with the requirements of the Engineer, afford all reasonable opportunities for carrying out their work to:

(a) any other contractors employed by the Employer and their workmen,

(b) the workmen of the Employer, and

(c) the workmen of any duly constituted authorities who may be employed in the execution on or near the Site of any work not included in the Contract or of any contract which the Employer may enter into in connection with or ancillary to the Works.

Facilities for Other Contractors 31.2 If, however, pursuant to Sub-Clause 31.1 the Contractor shall, on the written request of the Engineer:

(a) make available to any such other contractor, or to the Employer or any such authority, any roads or ways for the maintenance of which the Contractor is responsible, or

(b) permit the use, by any such, of Temporary Works or Contractor's Equipment on the Site, or

(c) provide any other service of whatsoever nature for any such, the Engineer shall determine an addition to the Contract Price in accordance with Clause 52 and shall notify the Contractor accordingly, with a copy to the Employer.

Contractor to Keep Site Clear 32.1 During the execution of the Works the Contractor shall keep the Site reasonably free from all unnecessary obstruction and shall store or dispose of any Contractor's Equipment and surplus materials and clear away and remove from the Site any wreckage, rubbish or Temporary Works no longer required.

Clearance of Site on Completion 33.1 Upon the issue of any Taking-Over Certificate the Contractor shall clear away and remove from that part of the Site to which such Taking-Over Certificate relates all Contractor's Equipment, surplus material, rubbish and Temporary Works of every kind, and leave such part of the Site and Works clean and in a workmanlike condition to the satisfaction of the Engineer. Provided that the Contractor shall be entitled to retain on Site, until the end of the Defects Liability Period, such materials, Contractor's Equipment and Temporary Works as are required by him for the purpose of fulfilling his obligations during the Defects Liability Period.

Labour

Engagement of Staff and Labour 34.1 The Contractor shall, unless otherwise provided in the Contract, make his own arrangements for the engagement of all staff and labour, local or other, and for their payment, housing, feeding and transport.

Returns of Labour and Contractor's Equipment 35.1 The Contractor shall, if required by the Engineer, deliver to the Engineer a return in detail, in such form and at such intervals as the Engineer may prescribe, showing the staff and the numbers of the several classes of labour from time to time employed by the Contractor on the Site and such information respecting Contractor's Equipment as the Engineer may require.

Materials, Plant and Workmanship

Quality of Materials, Plant and Workmanship 36.1 All materials, Plant and workmanship shall be

(a) of the respective kinds described in the Contract and in accordance with the Engineer's instructions, and

(b) subjected from time to time to such tests as the Engineer may require at the place of manufacture, fabrication or preparation, or on the Site or at such other place or places as may be specified in the Contract, or at all or any of such places.

The Contractor shall provide such assistance, labour, electricity, fuels, stores, apparatus and instruments as are normally required for examining, measuring and testing any materials or Plant and shall supply samples of materials, before incorporation in the Works, for testing as may be selected and required by the Engineer.

Cost of Samples 36.2 All samples shall be supplied by the Contractor at his own cost if the supply thereof is clearly intended by or provided for in the Contract.

Cost of Tests 36.3 The cost of making any test shall be borne by the Contractor if such test is

(a) clearly intended by or provided for in the Contract, or

(b) particularised in the Contract (in cases only of a test under load or of a test to ascertain whether the design of any finished or partially finished work is appropriate for the purposes which it was intended to fulfil) in sufficient detail to enable the Contractor to price or allow for the same in his Tender.

Cost of Tests not Provided for   **36.4**   If any test required by the Engineer which is

(a) not so intended by or provided for, or

(b) (in the cases above mentioned) not so particularised, or

(c) (though so intended or provided for) required by the Engineer to be carried out at any place other than the Site or the place of manufacture, fabrication or preparation of the materials or Plant tested,

shows the materials, Plant or workmanship not to be in accordance with the provisions of the Contract to the satisfaction of the Engineer, then the cost of such test shall be borne by the Contractor, but in any other case Sub-Clause 36.5 shall apply.

Engineer's Determination where Tests not Provided for   **36.5**   Where, pursuant to Sub-Clause 36.4, this Sub-Clause applies the Engineer shall, after due consultation with the Employer and the Contractor, determine:

(a) any extension of time to which the Contractor is entitled under Clause 44, and

(b) the amount of such costs, which shall be added to the Contract Price,

and shall notify the Contractor accordingly, with a copy to the Employer.

Inspection of Operations   **37.1**   The Engineer, and any person authorised by him, shall at all reasonable times have access to the Site and to all workshops and places where materials or Plant are being manufactured, fabricated or prepared for the Works and the Contractor shall afford every facility for and every assistance in obtaining the right to such access.

Inspection and Testing   **37.2**   The Engineer shall be entitled, during manufacture, fabrication or preparation to inspect and test the materials and Plant to be supplied under the Contract. If materials or Plant are being manufactured, fabricated or prepared in workshops or places other than those of the Contractor, the Contractor shall obtain permission for the Engineer to carry out such inspection and testing in those workshops or places. Such inspection or testing shall not release the Contractor from any obligation under the Contract.

Dates for Inspection and Testing   **37.3**   The Contractor shall agree with the Engineer on the time and place for the inspection or testing of any materials or Plant as provided in the Contract. The Engineer shall give the Contractor not less than 24 hours notice of his intention to carry out the inspection or to attend the tests. If the Engineer, or his duly authorised representative, does not attend on the date agreed, the Contractor may, unless otherwise instructed by the Engineer, proceed with the tests, which shall be deemed to have been made in the presence of the Engineer. The Contractor shall forthwith forward to the Engineer duly certified copies of the test readings. If the Engineer has not attended the tests, he shall accept the said readings as accurate.

Rejection   **37.4**   If, at the time and place agreed in accordance with Sub-Clause 37.3, the materials or Plant are not ready for inspection or testing or if, as a result of the inspection or testing referred to in this Clause, the Engineer determines that the materials or Plant are defective or otherwise not in accordance with the Contract, he may reject the materials or Plant and shall notify the Contractor thereof immediately. The notice shall state the Engineer's objections with reasons. The Contractor shall then promptly make good the defect or ensure that rejected materials or Plant comply with the Contract. If the Engineer so requests, the tests of rejected materials or Plant shall be made or repeated under the same terms and conditions. All costs incurred by the Employer by the repetition of the tests shall, after due consultation with the Employer and the Contractor, be determined by the Engineer and shall be recoverable from the Contractor by the Employer and may be deducted from any monies due or to become due to the Contractor and the Engineer shall notify the Contractor accordingly, with a copy to the Employer.

Independent Inspection 37.5 The Engineer may delegate inspection and testing of materials or Plant to an independent inspector. Any such delegation shall be effected in accordance with Sub-Clause 2.4 and for this purpose such independent inspector shall be considered as an assistant of the Engineer. Notice of such appointment (not being less than 14 days) shall be given by the Engineer to the Contractor.

Examination of Work before Covering up 38.1 No part of the Works shall be covered up or put out of view without the approval of the Engineer and the Contractor shall afford full opportunity for the Engineer to examine and measure any such part of the Works which is about to be covered up or put out of view and to examine foundations before any part of the Works is placed thereon. The Contractor shall give notice to the Engineer whenever any such part of the Works or foundations is or are ready or about to be ready for examination and the Engineer shall, without unreasonable delay, unless he considers it unnecessary and advises the Contractor accordingly, attend for the purpose of examining and measuring such part of the Works or of examining such foundations.

Uncovering and Making Openings 38.2 The Contractor shall uncover any part of the Works or make openings in or through the same as the Engineer may from time to time instruct and shall reinstate and make good such part. If any such part has been covered up or put out of view after compliance with the requirement of Sub-Clause 38.1 and is found to be executed in accordance with the Contract, the Engineer shall, after due consultation with the Employer and the Contractor, determine the amount of the Contractor's costs in respect of such of uncovering, making openings in or through, reinstating and making good the same, which shall be added to the Contract Price, and shall notify the Contractor accordingly, with a copy to the Employer. In any other case all costs shall be borne by the Contractor.

Removal of Improper Work, Materials or Plant 39.1 The Engineer shall have authority to issue instructions from time to time, for:

(a) the removal from the Site, within such time or times as may be specified in the instruction, of any materials or Plant which, in the opinion of the Engineer, are not in accordance with the Contract,

(b) the substitution of proper and suitable materials or Plant, and

(c) the removal and proper re-execution, notwithstanding any previous test thereof or interim payment therefor, of any work which, in respect of

(i) materials, Plant or workmanship, or

(ii) design by the Contractor or for which he is responsible,

is not, in the opinion of the Engineer, in accordance with the Contract.

Default of Contractor in Compliance 39.2 In case of default on the part of the Contractor in carrying out such instruction within the time specified therein or, if none, within a reasonable time, the Employer shall be entitled to employ and pay other persons to carry out the same and all costs consequent thereon or incidental thereto shall, after due consultation with the Employer and the Contractor, be determined by the Engineer and shall be recoverable from the Contractor by the Employer, and may be deducted by the Employer from any monies due or to become due to the Contractor and the Engineer shall notify the Contractor accordingly, with a copy to the Employer.

Suspension

Suspension of Work 40.1 The Contractor shall, on the instructions of the Engineer, suspend the progress of the Works or any part thereof for such time and in such manner as the Engineer may consider necessary and shall, during such suspension, properly protect and secure the Works or such part thereof so far as is necessary in the opinion of the Engineer. Unless such suspension is

(a) otherwise provided for in the Contract, or

(b) necessary by reason of some default of or breach of contract by the Contractor or for which he is responsible, or

(c) necessary by reason of climatic conditions on the Site, or

(d) necessary for the proper execution of the Works or for the safety of the Works or any part thereof (save to the extent that such necessity arises from any act or default by the Engineer or the Employer or from any of the risks defined in Sub-Clause 20.4),

Sub-Clause 40.2 shall apply.

Engineer's Determination following Suspension   **40.2**   Where, pursuant to Sub-Clause 40.1, this Sub-Clause applies the Engineer shall, after due consultation with the Employer and the Contractor, determine

(a) any extension of time to which the Contractor is entitled under Clause 44, and

(b) the amount, which shall be added to the Contract Price, in respect of the cost incurred by the Contractor by reason of such suspension,

and shall notify the Contractor accordingly, with a copy to the Employer.

Suspension lasting more than 84 Days   **40.3**   If the progress of the Works or any part thereof is suspended on the written instructions of the Engineer and if permission to resume work is not given by the Engineer within a period of 84 days from the date of suspension then, unless such suspension is within paragraph (a), (b), (c) or (d) of Sub-Clause 40.1, the Contractor may give notice to the Engineer requiring permission, within 28 days from the receipt thereof, to proceed with the Works or that part thereof in regard to which progress is suspended. If, within the said time, such permission is not granted, the Contractor may, but is not bound to, elect to treat the suspension, where it affects part only of the Works, as an omission of such part under Clause 51 by giving a further notice to the Engineer to that effect, or, where it affects the whole of the Works, treat the suspension as an event of default by the Employer and terminate his employment under the Contract in accordance with the provisions of Sub-Clause 69.1, whereupon the provisions of Sub-Clauses 69.2 and 69.3 shall apply.

Commencement and Delays

Commencement of Works   **41.1**   The Contractor shall commence the Works as soon as is reasonably possible after the receipt by him of a notice to this effect from the Engineer, which notice shall be issued within the time stated in the Appendix to Tender after the date of the Letter of Acceptance. Thereafter, the Contractor shall proceed with the Works with due expedition and without delay.

Possession of Site and Access Thereto   **42.1**   Save insofar as the Contract may prescribe:

(a) the extent of portions of the Site of which the Contractor is to be given possession from time to time, and

(b) the order in which such portions shall be made available to the Contractor

and subject to any requirement in the Contract as to the order in which the Works shall be executed, the Employer will, with the Engineer's notice to commence the Works, give to the Contractor possession of

(c) so much of the Site, and

(d) such access as, in accordance with the Contract, is to be provided by the Employer as may be required to enable the Contractor to commence and proceed with the execution of the Works in accordance with the programme referred to in Clause 14, if any, and otherwise in accordance with such reasonable proposals as the Contractor shall, by notice to the Engineer with a copy to the Employer, make. The Employer will, from time to time as the Works proceed, give to the Contractor possession of such further portions of the Site as may be required to enable the Contractor to proceed with the execution of the Works with due dispatch in accordance with such programme or proposals, as the case may be.

Failure to Give Possession 42.2 If the Contractor suffers delay and/or incurs costs from failure on the part of the Employer to give possession in accordance with the terms of Sub-Clause 42.1, the Engineer shall, after due consultation with the Employer and the Contractor, determine:

(a) any extension of time to which the Contractor is entitled under Clause 44, and

(b) the amount of such costs, which shall be added to the Contract Price,

and shall notify the Contractor accordingly, with a copy to the Employer.

Wayleaves and Facilities 42.3 The Contractor shall bear all costs and charges for special or temporary wayleaves required by him in connection with access to the Site. The Contractor shall also provide at his own cost any additional facilities outside the Site required by him for the purposes of the Works.

Time for Completion 43.1 The whole of the Works and, if applicable, any Section required to be completed within a particular time as stated in the Appendix to Tender, shall be completed, in accordance with the provisions of Clause 48, within the time stated in the Appendix to Tender for the whole of the Works or the Section (as the case may be), calculated from the Commencement Date, or such extended time as may be allowed under Clause 44.

Extension of Time for Completion 44.1 In the event of

(a) the amount or nature of extra or additional work, or

(b) any cause of delay referred to in these Conditions, or

(c) exceptionally adverse climatic conditions, or

(d) any delay, impediment or prevention by the Employer, or

(e) other special circumstances which may occur, other than through a default of or breach of contract by the Contractor or for which he is responsible,

being such as fairly to entitle the Contractor to an extension of the Time for Completion of the Works, or any Section or part thereof, the Engineer shall, after due consultation with the Employer and the Contractor, determine the amount of such extension and shall notify the Contractor accordingly, with a copy to the Employer.

Contractor to Provide Notification and Detailed Particulars 44.2 Provided that the Engineer is not bound to make any determination unless the Contractor has

(a) within 28 days after such event has first arisen notified the Engineer with a copy to the Employer, and

(b) within 28 days, or such other reasonable time as may be agreed by the Engineer, after such notification submitted to the Engineer detailed particulars of any extension of time to which he may consider himself entitled in order that such submission may be investigated at the time.

© FIDIC 1987

Interim Determination of Extension 44.3 Provided also that where an event has a continuing effect such that it is not practicable for the Contractor to submit detailed particulars within the period of 28 days referred to in Sub-Clause 44.2(b), he shall nevertheless be entitled to an extension of time provided that he has submitted to the Engineer interim particulars at intervals of not more than 28 days and final particulars within 28 days of the end of the effects resulting from the event. On receipt of such interim particulars, the Engineer shall, without undue delay, make an interim determination of extension of time and, on receipt of the final particulars, the Engineer shall review all the circumstances and shall determine an overall extension of time in regard to the event. In both such cases the Engineer shall notify the Contractor accordingly, with a copy to the Employer. No final review shall result in a decrease of any extension of time already determined by the Engineer.

Restriction on Working Hours 45.1 Subject to any provision to the contrary contained in the Contract, none of the Works shall, save as hereinafter provided, be carried on during the night or on locally recognised days of rest without the consent of the Engineer, except when work is unavoidable or absolutely necessary for the saving of life or property or for the safety of the Works, in which case the Contractor shall immediately advise the Engineer. Provided that the provisions of this Clause shall not be applicable in the case of any work which it is customary to carry out by multiple shifts.

Rate of Progress 46.1 If for any reason, which does not entitle the Contractor to an extension of time, the rate of progress of the Works or any Section is at any time, in the opinion of the Engineer, too slow to comply with the Time for Completion, the Engineer shall so notify the Contractor who shall thereupon take such steps as are necessary, subject to the consent of the Engineer, to expedite progress so as to comply with the Time for Completion. The Contractor shall not be entitled to any additional payment for taking such steps. If, as a result of any notice given by the Engineer under this Clause, the Contractor considers that it is necessary to do any work at night or on locally recognised days of rest, he shall be entitled to seek the consent of the Engineer so to do. Provided that if any steps, taken by the Contractor in meeting his obligations under this Clause, involve the Employer in additional supervision costs, such costs shall, after due consultation with the Employer and the Contractor, be determined by the Engineer and shall be recoverable from the Contractor by the Employer, and may be deducted by the Employer from any monies due or to become due to the Contractor and the Engineer shall notify the Contractor accordingly, with a copy to the Employer.

Liquidated Damages for Delay 47.1 If the Contractor fails to comply with the Time for Completion in accordance with Clause 48, for the whole of the Works or, if applicable, any Section within the relevant time prescribed by Clause 43, then the Contractor shall pay to the Employer the relevant sum stated in the Appendix to Tender as liquidated damages for such default and not as a penalty (which sum shall be the only monies due from the Contractor for such default) for every day or part of a day which shall elapse between the relevant Time for Completion and the date stated in a Taking-Over Certificate of the whole of the Works or the relevant Section, subject to the applicable limit stated in the Appendix to Tender. The Employer may, without prejudice to any other method of recovery, deduct the amount of such damages from any monies due or to become due to the Contractor. The payment or deduction of such damages shall not relieve the Contractor from his obligation to complete the Works, or from any other of his obligations and liabilities under the Contract.

Reduction of Liquidated Damages 47.2 If, before the Time for Completion of the whole of the Works or, if applicable, any Section, a Taking-Over Certificate has been issued for any part of the Works or of a Section, the liquidated damages for delay in completion of the remainder of the Works or of that Section shall, for any period of delay after the date stated in such Taking-Over Certificate, and in the absence of alternative provisions in the Contract, be reduced in the proportion which the value of the part so certified bears to the value of the whole of the Works or Section, as applicable. The provisions of this Sub-Clause shall only apply to the rate of liquidated damages and shall not affect the limit thereof.

| Taking-Over Certificate | 48.1 | When the whole of the Works have been substantially completed and have satisfactorily passed any Tests on Completion prescribed by the Contract, the Contractor may give a notice to that effect to the Engineer, with a copy to the Employer, accompanied by a written undertaking to finish with due expedition any outstanding work during the Defects Liability Period. Such notice and undertaking shall be deemed to be a request by the Contractor for the Engineer to issue a Taking-Over Certificate in respect of the Works. The Engineer shall, within 21 days of the date of delivery of such notice, either issue to the Contractor, with a copy to the Employer, a Taking-Over Certificate, stating the date on which, in his opinion, the Works were substantially completed in accordance with the Contract, or give instructions in writing to the Contractor specifying all the work which, in the Engineer's opinion, is required to be done by the Contractor before the issue of such Certificate. The Engineer shall also notify the Contractor of any defects in the Works affecting substantial completion that may appear after such instructions and before completion of the Works specified therein. The Contractor shall be entitled to receive such Taking-Over Certificate within 21 days of completion, to the satisfaction of the Engineer, of the Works so specified and remedying any defects so notified. |

| Taking Over of Sections or Parts | 48.2 | Similarly, in accordance with the procedure set out in Sub-Clause 48.1, the Contractor may request and the Engineer shall issue a Taking-Over Certificate in respect of: |

(a) any Section in respect of which a separate Time for Completion is provided in the Appendix to Tender, or

(b) any substantial part of the Permanent Works which has been both completed to the satisfaction of the Engineer and, otherwise than as provided for in the Contract, occupied or used by the Employer, or

(c) any part of the Permanent Works which the Employer has elected to occupy or use prior to completion (where such prior occupation or use is not provided for in the Contract or has not been agreed by the Contractor as a temporary measure).

| Substantial Completion of Parts | 48.3 | If any part of the Permanent Works has been substantially completed and has satisfactorily passed any Tests on Completion prescribed by the Contract, the Engineer may issue a Taking-Over Certificate in respect of that part of the Permanent Works before completion of the whole of the Works and, upon the issue of such Certificate, the Contractor shall be deemed to have undertaken to complete with due expedition any outstanding work in that part of the Permanent Works during the Defects Liability Period. |

| Surfaces Requiring Reinstatement | 48.4 | Provided that a Taking-Over Certificate given in respect of any Section or part of the Permanent Works before completion of the whole of the Works shall not be deemed to certify completion of any ground or surfaces requiring reinstatement, unless such Taking-Over Certificate shall expressly so state. |

Defects Liability

| Defects Liability Period | 49.1 | In these Conditions the expression "Defects Liability Period" shall mean the defects liability period named in the Appendix to Tender, calculated from: |

(a) the date of substantial completion of the Works certified by the Engineer in accordance with Clause 48, or

(b) in the event of more than one certificate having been issued by the Engineer under Clause 48, the respective dates so certified

and in relation to the Defects Liability Period the expression "the Works" shall be construed accordingly.

| Completion of Outstanding Work and Remedying Defects | 49.2 | To the intent that the Works shall, at or as soon as practicable after the expiration of the Defects Liability Period, be delivered to the Employer in the condition required by the Contract, fair wear and tear excepted, to the satisfaction of the Engineer, the Contractor shall: |

© FIDIC 1987

(a) complete the work, if any, outstanding on the date stated in the Taking-Over Certificate as soon as practicable after such date and

(b) execute all such work of amendment, reconstruction, and remedying defects, shrinkages or other faults as the Engineer may, during the Defects Liability Period or within 14 days after its expiration, as a result of an inspection made by or on behalf of the Engineer prior to its expiration, instruct the Contractor to execute.

Cost of Remedying Defects   **49.3**   All work referred to in Sub-Clause 49.2 (b) shall be executed by the Contractor at his own cost if the necessity thereof is, in the opinion of the Engineer, due to:

(a) the use of materials, Plant or workmanship not in accordance with the Contract, or

(b) where the Contractor is responsible for the design of part of the Permanent Works, any fault in such design, or

(c) the neglect or failure on the part of the Contractor to comply with any obligation, expressed or implied, on the Contractor's part under the Contract.

If, in the opinion of the Engineer, such necessity is due to any other cause, he shall determine an addition to the Contract Price in accordance with Clause 52 and shall notify the Contractor accordingly, with a copy to the Employer.

Contractor's Failure to Carry Out Instructions   **49.4**   In case of default on the part of the Contractor in carrying out such instruction within a reasonable time, the Employer shall be entitled to employ and pay other persons to carry out the same and if such work is work which, in the opinion of the Engineer, the Contractor was liable to do at his own cost under the Contract, then all costs consequent thereon or incidental thereto shall, after due consultation with the Employer and the Contractor, be determined by the Engineer and shall be recoverable from the Contractor by the Employer, and may be deducted by the Employer from any monies due or to become due to the Contractor and the Engineer shall notify the Contractor accordingly, with a copy to the Employer.

Contractor to Search   **50.1**   If any defect, shrinkage or other fault in the Works appears at any time prior to the end of the Defects Liability Period, the Engineer may instruct the Contractor, with copy to the Employer, to search under the directions of the Engineer for the cause thereof. Unless such defect, shrinkage or other fault is one for which the Contractor is liable under the Contract, the Engineer shall, after due consultation with the Employer and the Contractor, determine the amount in respect of the costs of such search incurred by the Contractor, which shall be added to the Contract Price and shall notify the Contractor accordingly, with a copy to the Employer. If such defect, shrinkage or other fault is one for which the Contractor is liable, the cost of the work carried out in searching as aforesaid shall be borne by the Contractor and he shall in such case remedy such defect, shrinkage or other fault at his own cost in accordance with the provisions of Clause 49.

Alterations, Additions and Omissions

Variations   **51.1**   The Engineer shall make any variation of the form, quality or quantity of the Works or any part thereof that may, in his opinion, be necessary and for that purpose, or if for any other reason it shall, in his opinion, be appropriate, he shall have the authority to instruct the Contractor to do and the Contractor shall do any of the following:

(a) increase or decrease the quantity of any work included in the Contract,

(b) omit any such work (but not if the omitted work is to be carried out by the Employer or by another contractor),

(c) change the character or quality or kind of any such work,

(d) change the levels, lines, position and dimensions of any part of the Works,

(e) execute additional work of any kind necessary for the completion of the Works,

(f) change any specified sequence or timing of construction of any part of the Works.

No such variation shall in any way vitiate or invalidate the Contract, but the effect, if any, of all such variations shall be valued in accordance with Clause 52. Provided that where the issue of an instruction to vary the Works is necessitated by some default of or breach of contract by the Contractor or for which he is responsible, any additional cost attributable to such default shall be borne by the Contractor.

Instructions for Variations | **51.2** | The Contractor shall not make any such variation without an instruction of the Engineer. Provided that no instruction shall be required for increase or decrease in the quantity of any work where such increase or decrease is not the result of an instruction given under this Clause, but is the result of the quantities exceeding or being less than those stated in the Bill of Quantities.

Valuation of Variations | **52.1** | All variations referred to in Clause 51 and any additions to the Contract Price which are required to be determined in accordance with Clause 52 (for the purposes of this Clause referred to as ''varied work''), shall be valued at the rates and prices set out in the Contract if, in the opinion of the Engineer, the same shall be applicable. If the Contract does not contain any rates or prices applicable to the varied work, the rates and prices in the Contract shall be used as the basis for valuation so far as may be reasonable, failing which, after due consultation by the Engineer with the Employer and the Contractor, suitable rates or prices shall be agreed upon between the Engineer and the Contractor. In the event of disagreement the Engineer shall fix such rates or prices as are, in his opinion, appropriate and shall notify the Contractor accordingly, with a copy to the Employer. Until such time as rates or prices are agreed or fixed, the Engineer shall determine provisional rates or prices to enable on-account payments to be included in certificates issued in accordance with Clause 60.

Power of Engineer to Fix Rates | **52.2** | Provided that if the nature or amount of any varied work relative to the nature or amount of the whole of the Works or to any part thereof, is such that, in the opinion of the Engineer, the rate or price contained in the Contract for any item of the Works is, by reason of such varied work, rendered inappropriate or inapplicable, then, after due consultation by the Engineer with the Employer and the Contractor, a suitable rate or price shall be agreed upon between the Engineer and the Contractor. In the event of disagreement the Engineer shall fix such other rate or price as is, in his opinion, appropriate and shall notify the Contractor accordingly, with a copy to the Employer. Until such time as rates or prices are agreed or fixed, the Engineer shall determine provisional rates or prices to enable on-account payments to be included in certificates issued in accordance with Clause 60.

Provided also that no varied work instructed to be done by the Engineer pursuant to Clause 51 shall be valued under Sub-Clause 52.1 or under this Sub-Clause unless, within 14 days of the date of such instruction and, other than in the case of omitted work, before the commencement of the varied work, notice shall have been given either:

(a) by the Contractor to the Engineer of his intention to claim extra payment or a varied rate or price, or

(b) by the Engineer to the Contractor of his intention to vary a rate or price.

Variations Exceeding 15 per cent | **52.3** | If, on the issue of the Taking-Over Certificate for the whole of the Works, it is found that as a result of:

(a) all varied work valued under Sub-Clauses 52.1 and 52.2, and

(b) all adjustments upon measurement of the estimated quantities set out in the Bill of Quantities, excluding Provisional Sums, dayworks and adjustments of price made under Clause 70,

© FIDIC 1987

but not from any other cause, there have been additions to or deductions from the Contract Price which taken together are in excess of 15 per cent of the "Effective Contract Price" (which for the purposes of this Sub-Clause shall mean the Contract Price, excluding Provisional Sums and allowance for dayworks, if any) then and in such event (subject to any action already taken under any other Sub-Clause of this Clause), after due consultation by the Engineer with the Employer and the Contractor, there shall be added to or deducted from the Contract Price such further sum as may be agreed between the Contractor and the Engineer or, failing agreement, determined by the Engineer having regard to the Contractor's Site and general overhead costs of the Contract. The Engineer shall notify the Contractor of any determination made under this Sub-Clause, with a copy to the Employer. Such sum shall be based only on the amount by which such additions or deductions shall be in excess of 15 per cent of the Effective Contract Price.

Daywork   **52.4**   The Engineer may, if in his opinion it is necessary or desirable, issue an instruction that any varied work shall be executed on a daywork basis. The Contractor shall then be paid for such varied work under the terms set out in the daywork schedule included in the Contract and at the rates and prices affixed thereto by him in the Tender.

The Contractor shall furnish to the Engineer such receipts or other vouchers as may be necessary to prove the amounts paid and, before ordering materials, shall submit to the Engineer quotations for the same for his approval.

In respect of such of the Works executed on a daywork basis, the Contractor shall, during the continuance of such work, deliver each day to the Engineer an exact list in duplicate of the names, occupation and time of all workmen employed on such work and a statement, also in duplicate, showing the description and quantity of all materials and Contractor's Equipment used thereon or therefor other than Contractor's Equipment which is included in the percentage addition in accordance with such daywork schedule. One copy of each list and statement will, if correct, or when agreed, be signed by the Engineer and returned to the Contractor.

At the end of each month the Contractor shall deliver to the Engineer a priced statement of the labour, materials and Contractor's Equipment, except as aforesaid, used and the Contractor shall not be entitled to any payment unless such lists and statements have been fully and punctually rendered. Provided always that if the Engineer considers that for any reason the sending of such lists or statements by the Contractor, in accordance with the foregoing provision, was impracticable he shall nevertheless be entitled to authorise payment for such work, either as daywork, on being satisfied as to the time employed and the labour, materials and Contractor's Equipment used on such work, or at such value therefor as shall, in his opinion, be fair and reasonable.

Procedure for Claims

Notice of Claims   **53.1**   Notwithstanding any other provision of the Contract, if the Contractor intends to claim any additional payment pursuant to any Clause of these Conditions or otherwise, he shall give notice of his intention to the Engineer, with a copy to the Employer, within 28 days after the event giving rise to the claim has first arisen.

Contemporary Records   **53.2**   Upon the happening of the event referred to in Sub-Clause 53.1, the Contractor shall keep such contemporary records as may reasonably be necessary to support any claim he may subsequently wish to make. Without necessarily admitting the Employer's liability, the Engineer shall, on receipt of a notice under Sub-Clause 53.1, inspect such contemporary records and may instruct the Contractor to keep any further contemporary records as are reasonable and may be material to the claim of which notice has been given. The Contractor shall permit the Engineer to inspect all records kept pursuant to this Sub-Clause and shall supply him with copies thereof as and when the Engineer so instructs.

**Substantiation
of Claims** 53.3 Within 28 days, or such other reasonable time as may be agreed by the Engineer, of giving notice under Sub-Clause 53.1, the Contractor shall send to the Engineer an account giving detailed particulars of the amount claimed and the grounds upon which the claim is based. Where the event giving rise to the claim has a continuing effect, such account shall be considered to be an interim account and the Contractor shall, at such intervals as the Engineer may reasonably require, send further interim accounts giving the accumulated amount of the claim and any further grounds upon which it is based. In cases where interim accounts are sent to the Engineer, the Contractor shall send a final account within 28 days of the end of the effects resulting from the event. The Contractor shall, if required by the Engineer so to do, copy to the Employer all accounts sent to the Engineer pursuant to this Sub-Clause.

Failure to Comply 53.4 If the Contractor fails to comply with any of the provisions of this Clause in respect of any claim which he seeks to make, his entitlement to payment in respect thereof shall not exceed such amount as the Engineer or any arbitrator or arbitrators appointed pursuant to Sub-Clause 67.3 assessing the claim considers to be verified by contemporary records (whether or not such records were brought to the Engineer's notice as required under Sub-Clauses 53.2 and 53.3).

Payment of Claims 53.5 The Contractor shall be entitled to have included in any interim payment certified by the Engineer pursuant to Clause 60 such amount in respect of any claim as the Engineer, after due consultation with the Employer and the Contractor, may consider due to the Contractor provided that the Contractor has supplied sufficient particulars to enable the Engineer to determine the amount due. If such particulars are insufficient to substantiate the whole of the claim, the Contractor shall be entitled to payment in respect of such part of the claim as such particulars may substantiate to the satisfaction of the Engineer. The Engineer shall notify the Contractor of any determination made under this Sub-Clause, with a copy to the Employer.

Contractor's Equipment, Temporary Works and Materials

**Contractor's
Equipment,
Temporary Works
and Materials;
Exclusive Use
for the Works** 54.1 All Contractor's Equipment, Temporary Works and materials provided by the Contractor shall, when brought on to the Site, be deemed to be exclusively intended for the execution of the Works and the Contractor shall not remove the same or any part thereof, except for the purpose of moving it from one part of the Site to another, without the consent of the Engineer. Provided that consent shall not be required for vehicles engaged in transporting any staff, labour, Contractor's Equipment, Temporary Works, Plant or materials to or from the Site.

**Employer not
Liable for Damage** 54.2 The Employer shall not at any time be liable, save as mentioned in Clauses 20 and 65, for the loss of or damage to any of the said Contractor's Equipment, Temporary Works or materials.

**Customs
Clearance** 54.3 The Employer will use his best endeavours in assisting the Contractor, where required, in obtaining clearance through the Customs of Contractor's Equipment, materials and other things required for the Works.

**Re-export of
Contractor's
Equipment** 54.4 In respect of any Contractor's Equipment which the Contractor has imported for the purposes of the Works, the Employer will use his best endeavours to assist the Contractor, where required, in procuring any necessary Government consent to the re-export of such Contractor's Equipment by the Contractor upon the removal thereof pursuant to the terms of the Contract.

Conditions of Hire of Contractor's Equipment 54.5 With a view to securing, in the event of termination under Clause 63, the continued availability, for the purpose of executing the Works, of any hired Contractor's Equipment, the Contractor shall not bring on to the Site any hired Contractor's Equipment unless there is an agreement for the hire thereof (which agreement shall be deemed not to include an agreement for hire purchase) which contains a provision that the owner thereof will, on request in writing made by the Employer within 7 days after the date on which any termination has become effective, and on the Employer undertaking to pay all hire charges in respect thereof from such date, hire such Contractor's Equipment to the Employer on the same terms in all respects as the same was hired to the Contractor save that the Employer shall be entitled to permit the use thereof by any other contractor employed by him for the purpose of executing and completing the Works and remedying any defects therein, under the terms of the said Clause 63.

Costs for the Purpose of Clause 63 54.6 In the event of the Employer entering into any agreement for the hire of Contractor's Equipment pursuant to Sub-Clause 54.5, all sums properly paid by the Employer under the provisions of any such agreement and all costs incurred by him (including stamp duties) in entering into such agreement shall be deemed, for the purpose of Clause 63, to be part of the cost of executing and completing the Works and the remedying of any defects therein.

Incorporation of Clause in Subcontracts 54.7 The Contractor shall, where entering into any subcontract for the execution of any part of the Works, incorporate in such subcontract (by reference or otherwise) the provisions of this Clause in relation to Contractor's Equipment, Temporary Works or materials brought on to the Site by the Subcontractor.

Approval of Materials not Implied 54.8 The operation of this Clause shall not be deemed to imply any approval by the Engineer of the materials or other matters referred to therein nor shall it prevent the rejection of any such materials at any time by the Engineer.

Measurement

Quantities 55.1 The quantities set out in the Bill of Quantities are the estimated quantities for the Works, and they are not to be taken as the actual and correct quantities of the Works to be executed by the Contractor in fulfilment of his obligations under the Contract.

Works to be Measured 56.1 The Engineer shall, except as otherwise stated, ascertain and determine by measurement the value of the Works in accordance with the Contract and the Contractor shall be paid that value in accordance with Clause 60. The Engineer shall, when he requires any part of the Works to be measured, give reasonable notice to the Contractor's authorised agent, who shall:

(a) forthwith attend or send a qualified representative to assist the Engineer in making such measurement, and

(b) supply all particulars required by the Engineer.

Should the Contractor not attend, or neglect or omit to send such representative, then the measurement made by the Engineer or approved by him shall be taken to be the correct measurement of such part of the Works. For the purpose of measuring such Permanent Works as are to be measured by records and drawings, the Engineer shall prepare records and drawings as the work proceeds and the Contractor, as and when called upon to do so in writing, shall, within 14 days, attend to examine and agree such records and drawings with the Engineer and shall sign the same when so agreed. If the Contractor does not attend to examine and agree such records and drawings, they shall be taken to be correct. If, after examination of such records and drawings, the Contractor does not agree the same or does not sign the same as agreed, they shall nevertheless be taken to be correct, unless the Contractor, within 14 days of such examination, lodges with the Engineer notice of the respects in which such records and drawings are claimed by him to be incorrect. On receipt of such notice, the Engineer shall review the records and drawings and either confirm or vary them.

© FIDIC 1987

Method of Measurement	57.1	The Works shall be measured net, notwithstanding any general or local custom, except where otherwise provided for in the Contract.
Breakdown of Lump Sum Items	57.2	For the purposes of statements submitted in accordance with Sub-Clause 60.1, the Contractor shall submit to the Engineer, within 28 days after the receipt of the Letter of Acceptance, a breakdown for each of the lump sum items contained in the Tender. Such breakdowns shall be subject to the approval of the Engineer.

Provisional Sums

Definition of "Provisional Sum"	58.1	"Provisional Sum" means a sum included in the Contract and so designated in the Bill of Quantities for the execution of any part of the Works or for the supply of goods, materials, Plant or services, or for contingencies, which sum may be used, in whole or in part, or not at all, on the instructions of the Engineer. The Contractor shall be entitled to only such amounts in respect of the work, supply or contingencies to which such Provisional Sums relate as the Engineer shall determine in accordance with this Clause. The Engineer shall notify the Contractor of any determination made under this Sub-Clause, with a copy to the Employer.
Use of Provisional Sums	58.2	In respect of every Provisional Sum the Engineer shall have authority to issue instructions for the execution of work or for the supply of goods, materials, Plant or services by:

(a) the Contractor, in which case the Contractor shall be entitled to an amount equal to the value thereof determined in accordance with Clause 52,

(b) a nominated Subcontractor, as hereinafter defined, in which case the sum to be paid to the Contractor therefor shall be determined and paid in accordance with Sub-Clause 59.4.

Production of Vouchers	58.3	The Contractor shall produce to the Engineer all quotations, invoices, vouchers and accounts or receipts in connection with expenditure in respect of Provisional Sums, except where work is valued in accordance with rates or prices set out in the Tender.

Nominated Subcontractors

Definition of "Nominated Subcontractors"	59.1	All specialists, merchants, tradesmen and others executing any work or supplying any goods, materials, Plant or services for which Provisional Sums are included in the Contract, who may have been or be nominated or selected or approved by the Employer or the Engineer, and all persons to whom by virtue of the provisions of the Contract the Contractor is required to subcontract shall, in the execution of such work or the supply of such goods, materials, Plant or services, be deemed to be subcontractors to the Contractor and are referred to in this Contract as "nominated Subcontractors".
Nominated Subcontractors; Objection to Nomination	59.2	The Contractor shall not be required by the Employer or the Engineer, or be deemed to be under any obligation, to employ any nominated Subcontractor against whom the Contractor may raise reasonable objection, or who declines to enter into a subcontract with the Contractor containing provisions:

(a) that in respect of the work, goods, materials, Plant or services the subject of the subcontract, the nominated Subcontractor will undertake towards the Contractor such obligations and liabilities as will enable the Contractor to discharge his own obligations and liabilities towards the Employer under the terms of the Contract and will save harmless and indemnify the Contractor from and against the same and from all claims, proceedings, damages, costs, charges and expenses whatsoever arising out of or in connection therewith, or arising out of or in connection with any failure to perform such obligations or to fulfil such liabilities, and

(b) that the nominated Subcontractor will save harmless and indemnify the Contractor from and against any negligence by the nominated Subcontractor, his agents, workmen and servants and from and against any misuse by him or them of any Temporary Works provided by the Contractor for the purposes of the Contract and from all claims as aforesaid.

Design Requirements to be Expressly Stated 59.3 If in connection with any Provisional Sum the services to be provided include any matter of design or specification of any part of the Permanent Works or of any Plant to be incorporated therein, such requirement shall be expressly stated in the Contract and shall be included in any nominated Subcontract. The nominated Subcontract shall specify that the nominated Subcontractor providing such services will save harmless and indemnify the Contractor from and against the same and from all claims, proceedings, damages, costs, charges and expenses whatsoever arising out of or in connection with any failure to perform such obligations or to fulfil such liabilities.

Payments to Nominated Subcontractors 59.4 For all work executed or goods, materials, Plant or services supplied by any nominated Subcontractor, the Contractor shall be entitled to:

(a) the actual price paid or due to be paid by the Contractor, on the instructions of the Engineer, and in accordance with the subcontract;

(b) in respect of labour supplied by the Contractor, the sum, if any, entered in the Bill of Quantities or, if instructed by the Engineer pursuant to paragraph (a) of Sub-Clause 58.2, as may be determined in accordance with Clause 52;

(c) in respect of all other charges and profit, a sum being a percentage rate of the actual price paid or due to be paid calculated, where provision has been made in the Bill of Quantities for a rate to be set against the relevant Provisional Sum, at the rate inserted by the Contractor against that item or, where no such provision has been made, at the rate inserted by the Contractor in the Appendix to Tender and repeated where provision for such is made in a special item provided in the Bill of Quantities for such purpose.

Certification of Payments to Nominated Subcontractors 59.5 Before issuing, under Clause 60, any certificate, which includes any payment in respect of work done or goods, materials, Plant or services supplied by any nominated Subcontractor, the Engineer shall be entitled to demand from the Contractor reasonable proof that all payments, less retentions, included in previous certificates in respect of the work or goods, materials, Plant or services of such nominated Subcontractor have been paid or discharged by the Contractor. If the Contractor fails to supply such proof then, unless the Contractor

(a) satisfies the Engineer in writing that he has reasonable cause for withholding or refusing to make such payments and

(b) produces to the Engineer reasonable proof that he has so informed such nominated Subcontractor in writing,

the Employer shall be entitled to pay to such nominated Subcontractor direct, upon the certificate of the Engineer, all payments, less retentions, provided for in the nominated Subcontract, which the Contractor has failed to make to such nominated Subcontractor and to deduct by way of set-off the amount so paid by the Employer from any sums due or to become due from the Employer to the Contractor.

Provided that, where the Engineer has certified and the Employer has paid direct as aforesaid, the Engineer shall, in issuing any further certificate in favour of the Contractor, deduct from the amount thereof the amount so paid, direct as aforesaid, but shall not withhold or delay the issue of the certificate itself when due to be issued under the terms of the Contract.

Certificates and Payment

Monthly **60.1** The Contractor shall submit to the Engineer after the end of each month six
Statements copies, each signed by the Contractor's representative approved by the Engineer
in accordance with Sub-Clause 15.1, of a statement, in such form as the Engineer
may from time to time prescribe, showing the amounts to which the Contractor
considers himself to be entitled up to the end of the month in respect of

(a) the value of the Permanent Works executed

(b) any other items in the Bill of Quantities including those for Contractor's
Equipment, Temporary Works, dayworks and the like

(c) the percentage of the invoice value of listed materials, all as stated in the
Appendix to Tender, and Plant delivered by the Contractor on the Site for
incorporation in the Permanent Works but not incorporated in such Works

(d) adjustments under Clause 70

(e) any other sum to which the Contractor may be entitled under the Contract.

Monthly Payments **60.2** The Engineer shall, within 28 days of receiving such statement, certify to the
Employer the amount of payment to the Contractor which he considers due and
payable in respect thereof, subject:

(a) firstly, to the retention of the amount calculated by applying the Percentage of
Retention stated in the Appendix to Tender, to the amount to which the
Contractor is entitled under paragraphs (a), (b), (c) and (e) of Sub-Clause 60.1
until the amount so retained reaches the Limit of Retention Money stated in the
Appendix to Tender, and

(b) secondly, to the deduction, other than pursuant to Clause 47, of any sums
which may have become due and payable by the Contractor to the Employer.

Provided that the Engineer shall not be bound to certify any payment under this
Sub-Clause if the net amount thereof, after all retentions and deductions, would
be less than the Minimum Amount of Interim Certificates stated in the Appendix
to Tender.

Notwithstanding the terms of this Clause or any other Clause of the Contract no
amount will be certified by the Engineer for payment until the performance
security, if required under the Contract, has been provided by the Contractor and
approved by the Employer.

Payment of **60.3** (a) Upon the issue of the Taking-Over Certificate with respect to the whole of the
Retention Money Works, one half of the Retention Money, or upon the issue of a Taking-Over
Certificate with respect to a Section or part of the Permanent Works only such
proportion thereof as the Engineer determines having regard to the relative value
of such Section or part of the Permanent Works, shall be certified by the Engineer
for payment to the Contractor.

(b) Upon the expiration of the Defects Liability Period for the Works the other
half of the Retention Money shall be certified by the Engineer for payment to the
Contractor. Provided that, in the event of different Defects Liability Periods
having become applicable to different Sections or parts of the Permanent Works
pursuant to Clause 48, the expression "expiration of the Defects Liability
Period" shall, for the purposes of this Sub-Clause, be deemed to mean the
expiration of the latest of such periods.

Provided also that if at such time, there shall remain to be executed by the
Contractor any work ordered, pursuant to Clauses 49 and 50, in respect of the
Works, the Engineer shall be entitled to withhold certification until completion of
such work of so much of the balance of the Retention Money as shall, in the
opinion of the Engineer, represent the cost of the work remaining to be executed.

Correction of Certificates	**60.4**	The Engineer may by any interim certificate make any correction or modification in any previous certificate which shall have been issued by him and shall have authority, if any work is not being carried out to his satisfaction, to omit or reduce the value of such work in any interim certificate.

Statement at Completion   **60.5**   Not later than 84 days after the issue of the Taking-Over Certificate in respect of the whole of the Works, the Contractor shall submit to the Engineer a Statement at Completion with supporting documents showing in detail, in the form approved by the Engineer,

(a) the final value of all work done in accordance with the Contract up to the date stated in such Taking-Over Certificate

(b) any further sums which the Contractor considers to be due and

(c) an estimate of amounts which the Contractor considers will become due to him under the Contract.

Estimated amounts shall be shown separately in such Statement at Completion. The Engineer shall certify payment in accordance with Sub-Clause 60.2.

Final Statement   **60.6**   Not later than 56 days after the issue of the Defects Liability Certificate pursuant to Sub-Clause 62.1, the Contractor shall submit to the Engineer for consideration a draft final statement with supporting documents showing in detail, in the form approved by the Engineer,

(a) the value of all work done in accordance with the Contract and

(b) any further sums which the Contractor considers to be due to him under the Contract.

If the Engineer disagrees with or cannot verify any part of the draft final statement, the Contractor shall submit such further information as the Engineer may reasonably require and shall make such changes in the draft as may be agreed between them. The Contractor shall then prepare and submit to the Engineer the final statement as agreed (for the purposes of these Conditions referred to as the "Final Statement").

Discharge   **60.7**   Upon submission of the Final Statement, the Contractor shall give to the Employer, with a copy to the Engineer, a written discharge confirming that the total of the Final Statement represents full and final settlement of all monies due to the Contractor arising out of or in respect of the Contract. Provided that such discharge shall become effective only after payment due under the Final Certificate issued pursuant to Sub-Clause 60.8 has been made and the performance security referred to in Sub-Clause 10.1, if any, has been returned to the Contractor.

Final Certificate   **60.8**   Within 28 days after receipt of the Final Statement, and the written discharge, the Engineer shall issue to the Employer (with a copy to the Contractor) a Final Certificate stating

(a) the amount which, in the opinion of the Engineer, is finally due under the Contract, and

(b) after giving credit to the Employer for all amounts previously paid by the Employer and for all sums to which the Employer is entitled under the Contract, other than Clause 47, the balance, if any, due from the Employer to the Contractor or from the Contractor to the Employer as the case may be.

Cessation of Employer's Liability   **60.9**   The Employer shall not be liable to the Contractor for any matter or thing arising out of or in connection with the Contract or execution of the Works, unless the Contractor shall have included a claim in respect thereof in his Final Statement and (except in respect of matters or things arising after the issue of the Taking-Over Certificate in respect of the whole of the Works) in the Statement at Completion referred to in Sub-Clause 60.5.

Time for Payment	**60.10**	The amount due to the Contractor under any interim certificate issued by the Engineer pursuant to this Clause, or to any other term of the Contract, shall, subject to Clause 47, be paid by the Employer to the Contractor within 28 days after such interim certificate has been delivered to the Employer, or, in the case of the Final Certificate referred to in Sub-Clause 60.8, within 56 days, after such Final Certificate has been delivered to the Employer. In the event of the failure of the Employer to make payment within the times stated, the Employer shall pay to the Contractor interest at the rate stated in the Appendix to Tender upon all sums unpaid from the date by which the same should have been paid. The provisions of this Sub-Clause are without prejudice to the Contractor's entitlement under Clause 69.
Approval only by Defects Liability Certificate	**61.1**	Only the Defects Liability Certificate, referred to in Clause 62, shall be deemed to constitute approval of the Works.
Defects Liability Certificate	**62.1**	The Contract shall not be considered as completed until a Defects Liability Certificate shall have been signed by the Engineer and delivered to the Employer, with a copy to the Contractor, stating the date on which the Contractor shall have completed his obligations to execute and complete the Works and remedy any defects therein to the Engineer's satisfaction. The Defects Liability Certificate shall be given by the Engineer within 28 days after the expiration of the Defects Liability Period, or, if different defects liability periods shall become applicable to different Sections or parts of the Permanent Works, the expiration of the latest such period, or as soon thereafter as any works instructed, pursuant to Clauses 49 and 50, have been completed to the satisfaction of the Engineer. Provided that the issue of the Defects Liability Certificate shall not be a condition precedent to payment to the Contractor of the second portion of the Retention Money in accordance with the conditions set out in Sub-Clause 60.3.
Unfulfilled Obligations	**62.2**	Notwithstanding the issue of the Defects Liability Certificate the Contractor and the Employer shall remain liable for the fulfilment of any obligation incurred under the provisions of the Contract prior to the issue of the Defects Liability Certificate which remains unperformed at the time such Defects Liability Certificate is issued and, for the purposes of determining the nature and extent of any such obligation, the Contract shall be deemed to remain in force between the parties to the Contract.

Remedies

Default of Contractor	**63.1**	If the Contractor is deemed by law unable to pay his debts as they fall due, or enters into voluntary or involuntary bankruptcy, liquidation or dissolution (other than a voluntary liquidation for the purposes of amalgamation or reconstruction), or becomes insolvent, or makes an arrangement with, or assignment in favour of, his creditors, or agrees to carry out the Contract under a committee of inspection of his creditors, or if a receiver, administrator, trustee or liquidator is appointed over any substantial part of his assets, or if, under any law or regulation relating to reorganization, arrangement or readjustment of debts, proceedings are commenced against the Contractor or resolutions passed in connection with dissolution or liquidation or if any steps are taken to enforce any security interest over a substantial part of the assets of the Contractor, or if any act is done or event occurs with respect to the Contractor or his assets which, under any applicable law has a substantially similar effect to any of the foregoing acts or events, or if the Contractor has contravened Sub-Clause 3.1, or has an execution levied on his goods, or if the Engineer certifies to the Employer, with a copy to the Contractor, that, in his opinion, the Contractor:

(a) has repudiated the Contract, or

(b) without reasonable excuse has failed
    (i) to commence the Works in accordance with Sub-Clause 41.1, or

(ii) to proceed with the Works, or any Section thereof, within 28 days after receiving notice pursuant to Sub-Clause 46.1, or

(c) has failed to comply with a notice issued pursuant to Sub-Clause 37.4 or an instruction issued pursuant to Sub-Clause 39.1 within 28 days after having received it, or

(d) despite previous warning from the Engineer, in writing, is otherwise persistently or flagrantly neglecting to comply with any of his obligations under the Contract, or

(e) has contravened Sub-Clause 4.1,

then the Employer may, after giving 14 days' notice to the Contractor, enter upon the Site and the Works and terminate the employment of the Contractor without thereby releasing the Contractor from any of his obligations or liabilities under the Contract, or affecting the rights and authorities conferred on the Employer or the Engineer by the Contract, and may himself complete the Works or may employ any other contractor to complete the Works. The Employer or such other contractor may use for such completion so much of the Contractor's Equipment, Temporary Works and materials as he or they may think proper.

Valuation at Date of Termination 63.2 The Engineer shall, as soon as may be practicable after any such entry and termination by the Employer, fix and determine ex parte, or by or after reference to the parties or after such investigation or enquiries as he may think fit to make or institute, and shall certify:

(a) what amount (if any) had, at the time of such entry and termination, been reasonably earned by or would reasonably accrue to the Contractor in respect of work then actually done by him under the Contract, and

(b) the value of any of the said unused or partially used materials, any Contractor's Equipment and any Temporary Works.

Payment after Termination 63.3 If the Employer terminates the Contractor's employment under this Clause, he shall not be liable to pay to the Contractor any further amount (including damages) in respect of the Contract until the expiration of the Defects Liability Period and thereafter until the costs of execution, completion and remedying of any defects, damages for delay in completion (if any) and all other expenses incurred by the Employer have been ascertained and the amount thereof certified by the Engineer. The Contractor shall then be entitled to receive only such sum (if any) as the Engineer may certify would have been payable to him upon due completion by him after deducting the said amount. If such amount exceeds the sum which would have been payable to the Contractor on due completion by him, then the Contractor shall, upon demand, pay to the Employer the amount of such excess and it shall be deemed a debt due by the Contractor to the Employer and shall be recoverable accordingly.

Assignment of Benefit of Agreement 63.4 Unless prohibited by law, the Contractor shall, if so instructed by the Engineer within 14 days of such entry and termination referred to in Sub-Clause 63.1, assign to the Employer the benefit of any agreement for the supply of any goods or materials or services and/or for the execution of any work for the purposes of the Contract, which the Contractor may have entered into.

Urgent Remedial Work	64.1	If, by reason of any accident, or failure, or other event occurring to, in, or in connection with the Works, or any part thereof, either during the execution of the Works, or during the Defects Liability Period, any remedial or other work is, in the opinion of the Engineer, urgently necessary for the safety of the Works and the Contractor is unable or unwilling at once to do such work, the Employer shall be entitled to employ and pay other persons to carry out such work as the Engineer may consider necessary. If the work or repair so done by the Employer is work which, in the opinion of the Engineer, the Contractor was liable to do at his own cost under the Contract, then all costs consequent thereon or incidental thereto shall, after due consultation with the Employer and the Contractor, be determined by the Engineer and shall be recoverable from the Contractor by the Employer, and may be deducted by the Employer from any monies due or to become due to the Contractor and the Engineer shall notify the Contractor accordingly, with a copy to the Employer. Provided that the Engineer shall, as soon after the occurrence of any such emergency as may be reasonably practicable, notify the Contractor thereof.

Special Risks

No Liability for Special Risks	65.1	The Contractor shall be under no liability whatsoever in consequence of any of the special risks referred to in Sub-Clause 65.2, whether by way of indemnity or otherwise, for or in respect of:

(a) destruction of or damage to the Works, save to work condemned under the provisions of Clause 39 prior to the occurrence of any of the said special risks, or

(b) destruction of or damage to property, whether of the Employer or third parties, or

(c) injury or loss of life.

Special Risks	65.2	The special risks are:

(a) the risks defined under paragraphs (a), (c), (d) and (e) of Sub-Clause 20.4, and

(b) the risks defined under paragraph (b) of Sub-Clause 20.4 insofar as these relate to the country in which the Works are to be executed.

Damage to Works by Special Risks	65.3	If the Works or any materials or Plant on or near or in transit to the Site, or any of the Contractor's Equipment, sustain destruction or damage by reason of any of the said special risks, the Contractor shall be entitled to payment in accordance with the Contract for any Permanent Works duly executed and for any materials or Plant so destroyed or damaged and, so far as may be required by the Engineer or as may be necessary for the completion of the Works, to payment for:

(a) rectifying any such destruction or damage to the Works, and

(b) replacing or rectifying such materials or Contractor's Equipment

and the Engineer shall determine an addition to the Contract Price in accordance with Clause 52 (which shall in the case of the cost of replacement of Contractor's Equipment include the fair market value thereof as determined by the Engineer) and shall notify the Contractor accordingly, with a copy to the Employer.

Projectile, Missile	65.4	Destruction, damage, injury or loss of life caused by the explosion or impact, whenever and wherever occurring, of any mine, bomb, shell, grenade, or other projectile, missile, munition, or explosive of war, shall be deemed to be a consequence of the said special risks.

Increased Costs arising from Special Risks	65.5	Save to the extent that the Contractor is entitled to payment under any other provision of the Contract, the Employer shall repay to the Contractor any costs of the execution of the Works (other than such as may be attributable to the cost of reconstructing work condemned under the provisions of Clause 39 prior to the occurrence of any special risk) which are howsoever attributable to or consequent on or the result of or in any way whatsoever connected with the said special risks, subject however to the provisions in this Clause hereinafter contained in regard to outbreak of war, but the Contractor shall, as soon as any such cost comes to his knowledge, forthwith notify the Engineer thereof. The Engineer shall, after due consultation with the Employer and the Contractor, determine the amount of the Contractor's costs in respect thereof which shall be added to the Contract Price and shall notify the Contractor accordingly, with a copy to the Employer.
Outbreak of War	65.6	If, during the currency of the Contract, there is an outbreak of war, whether war is declared or not, in any part of the world which, whether financially or otherwise, materially affects the execution of the Works, the Contractor shall, unless and until the Contract is terminated under the provisions of this Clause, continue to use his best endeavours to complete the execution of the Works. Provided that the Employer shall be entitled, at any time after such outbreak of war, to terminate the Contract by giving notice to the Contractor and, upon such notice being given, the Contract shall, except as to the rights of the parties under this Clause and to the operation of Clause 67, terminate, but without prejudice to the rights of either party in respect of any antecedent breach thereof.
Removal of Contractor's Equipment on Termination	65.7	If the Contract is terminated under the provisions of Sub-Clause 65.6, the Contractor shall, with all reasonable dispatch, remove from the Site all Contractor's Equipment and shall give similar facilities to his Subcontractors to do so.
Payment if Contract Terminated	65.8	If the Contract is terminated as aforesaid, the Contractor shall be paid by the Employer, insofar as such amounts or items have not already been covered by payments on account made to the Contractor, for all work executed prior to the date of termination at the rates and prices provided in the Contract and in addition:

(a) The amounts payable in respect of any preliminary items referred to in the Bill of Quantities, so far as the work or service comprised therein has been carried out or performed, and a proper proportion of any such items which have been partially carried out or performed.

(b) The cost of materials, Plant or goods reasonably ordered for the Works which have been delivered to the Contractor or of which the Contractor is legally liable to accept delivery, such materials, Plant or goods becoming the property of the Employer upon such payments being made by him.

(c) A sum being the amount of any expenditure reasonably incurred by the Contractor in the expectation of completing the whole of the Works insofar as such expenditure has not been covered by any other payments referred to in this Sub-Clause.

(d) Any additional sum payable under the provisions of Sub-Clauses 65.3 and 65.5.

(e) Such proportion of the cost as may be reasonable, taking into account payments made or to be made for work executed, of removal of Contractor's Equipment under Sub-Clause 65.7 and, if required by the Contractor, return thereof to the Contractor's main plant yard in his country of registration or to other destination, at no greater cost.

(f) The reasonable cost of repatriation of all the Contractor's staff and workmen employed on or in connection with the Works at the time of such termination.

Provided that against any payment due from the Employer under this Sub-Clause, the Employer shall be entitled to be credited with any outstanding balances due from the Contractor for advances in respect of Contractor's Equipment, materials and Plant and any other sums which, at the date of termination, were recoverable by the Employer from the Contractor under the terms of the Contract. Any sums payable under this Sub-Clause shall, after due consultation with the Employer and the Contractor, be determined by the Engineer who shall notify the Contractor accordingly, with a copy to the Employer.

Release from Performance

Payment in Event of Release from Performance

66.1 If any circumstance outside the control of both parties arises after the issue of the Letter of Acceptance which renders it impossible or unlawful for either party to fulfil his contractual obligations, or under the law governing the Contract the parties are released from further performance, then the sum payable by the Employer to the Contractor in respect of the work executed shall be the same as that which would have been payable under Clause 65 if the Contract had been terminated under the provisions of Clause 65.

Settlement of Disputes

Engineer's Decision

67.1 If a dispute of any kind whatsoever arises between the Employer and the Contractor in connection with, or arising out of, the Contract or the execution of the Works, whether during the execution of the Works or after their completion and whether before or after repudiation or other termination of the Contract, including any dispute as to any opinion, instruction, determination, certificate or valuation of the Engineer, the matter in dispute shall, in the first place, be referred in writing to the Engineer, with a copy to the other party. Such reference shall state that it is made pursuant to this Clause. No later than the eighty-fourth day after the day on which he received such reference the Engineer shall give notice of his decision to the Employer and the Contractor. Such decision shall state that it is made pursuant to this Clause.

Unless the Contract has already been repudiated or terminated, the Contractor shall, in every case, continue to proceed with the Works with all due diligence and the Contractor and the Employer shall give effect forthwith to every such decision of the Engineer unless and until the same shall be revised, as hereinafter provided, in an amicable settlement or an arbitral award.

If either the Employer or the Contractor be dissatisfied with any decision of the Engineer, or if the Engineer fails to give notice of his decision on or before the eighty-fourth day after the day on which he received the reference, then either the Employer or the Contractor may, on or before the seventieth day after the day on which he received notice of such decision, or on or before the seventieth day after the day on which the said period of 84 days expired, as the case may be, give notice to the other party, with a copy for information to the Engineer, of his intention to commence arbitration, as hereinafter provided as to the matter in dispute. Such notice shall establish the entitlement of the party giving the same to commence arbitration, as hereinafter provided, as to such dispute and, subject to Sub-Clause 67.4, no arbitration in respect thereof may be commenced unless such notice is given.

If the Engineer has given notice of his decision as to a matter in dispute to the Employer and the Contractor and no notification of intention to commence arbitration as to such dispute has been given by either the Employer or the Contractor on or before the seventieth day after the day on which the parties received notice as to such decision from the Engineer, the said decision shall become final and binding upon the Employer and the Contractor.

© FIDIC 1987

Amicable Settlement 67.2 Where notice of intention to commence arbitration as to a dispute has been given in accordance with Sub-Clause 67.1, arbitration of such dispute shall not be commenced unless an attempt has first been made by the parties to settle such dispute amicably. Provided that, unless the parties otherwise agree, arbitration may be commenced on or after the fifty-sixth day after the day on which notice of intention to commence arbitration of such dispute was given, whether or not any attempt at amicable settlement thereof has been made.

Arbitration 67.3 Any dispute in respect of which:

(a) the decision, if any, of the Engineer has not become final and binding pursuant to Sub-Clause 67.1, and

(b) amicable settlement has not been reached within the period stated in Sub-Clause 67.2

shall be finally settled, unless otherwise specified in the Contract, under the Rules of Conciliation and Arbitration of the International Chamber of Commerce by one or more arbitrators appointed under such Rules. The said arbitrator/s shall have full power to open up, review and revise any decision, opinion, instruction, determination, certificate or valuation of the Engineer related to the dispute.

Neither party shall be limited in the proceedings before such arbitrator/s to the evidence or arguments put before the Engineer for the purpose of obtaining his said decision pursuant to Sub-Clause 67.1. No such decision shall disqualify the Engineer from being called as a witness and giving evidence before the arbitrator/s on any matter whatsoever relevant to the dispute.

Arbitration may be commenced prior to or after completion of the Works, provided that the obligations of the Employer, the Engineer and the Contractor shall not be altered by reason of the arbitration being conducted during the progress of the Works.

Failure to Comply with Engineer's Decision 67.4 Where neither the Employer nor the Contractor has given notice of intention to commence arbitration of a dispute within the period stated in Sub-Clause 67.1 and the related decision has become final and binding, either party may, if the other party fails to comply with such decision, and without prejudice to any other rights it may have, refer the failure to arbitration in accordance with Sub-Clause 67.3. The provisions of Sub-Clauses 67.1 and 67.2 shall not apply to any such reference.

Notices

Notice to Contractor 68.1 All certificates, notices or instructions to be given to the Contractor by the Employer or the Engineer under the terms of the Contract shall be sent by post, cable, telex or facsimile transmission to or left at the Contractor's principal place of business or such other address as the Contractor shall nominate for that purpose.

Notice to Employer and Engineer 68.2 Any notice to be given to the Employer or to the Engineer under the terms of the Contract shall be sent by post, cable, telex or facsimile transmission to or left at the respective addresses nominated for that purpose in Part II of these Conditions.

Change of Address 68.3 Either party may change a nominated address to another address in the country where the Works are being executed by prior notice to the other party, with a copy to the Engineer, and the Engineer may do so by prior notice to both parties.

Default of Employer

Default of Employer **69.1** In the event of the Employer:

(a) failing to pay to the Contractor the amount due under any certificate of the Engineer within 28 days after the expiry of the time stated in Sub-Clause 60.10 within which payment is to be made, subject to any deduction that the Employer is entitled to make under the Contract, or

(b) interfering with or obstructing or refusing any required approval to the issue of any such certificate, or

(c) becoming bankrupt or, being a company, going into liquidation, other than for the purpose of a scheme of reconstruction or amalgamation, or

(d) giving notice to the Contractor that for unforeseen reasons, due to economic dislocation, it is impossible for him to continue to meet his contractual obligations

the Contractor shall be entitled to terminate his employment under the Contract by giving notice to the Employer, with a copy to the Engineer. Such termination shall take effect 14 days after the giving of the notice.

Removal of Contractor's Equipment **69.2** Upon the expiry of the 14 days' notice referred to in Sub-Clause 69.1, the Contractor shall, notwithstanding the provisions of Sub-Clause 54.1, with all reasonable despatch, remove from the Site all Contractor's Equipment brought by him thereon.

Payment on Termination **69.3** In the event of such termination the Employer shall be under the same obligations to the Contractor in regard to payment as if the Contract had been terminated under the provisions of Clause 65, but, in addition to the payments specified in Sub-Clause 65.8, the Employer shall pay to the Contractor the amount of any loss or damage to the Contractor arising out of or in connection with or by consequence of such termination.

Contractor's Entitlement to Suspend Work **69.4** Without prejudice to the Contractor's entitlement to interest under Sub-Clause 60.10 and to terminate under Sub-Clause 69.1, the Contractor may, if the Employer fails to pay the Contractor the amount due under any certificate of the Engineer within 28 days after the expiry of the time stated in Sub-Clause 60.10 within which payment is to be made, subject to any deduction that the Employer is entitled to make under the Contract, after giving 28 days' prior notice to the Employer, with a copy to the Engineer, suspend work or reduce the rate of work.

If the Contractor suspends work or reduces the rate of work in accordance with the provisions of this Sub-Clause and thereby suffers delay or incurs cost the Engineer shall, after due consultation with the Employer and the Contractor, determine

(a) any extension of time to which the Contractor is entitled under Clause 44, and

(b) the amount of such costs, which shall be added to the Contract Price,

and shall notify the Contractor accordingly, with a copy to the Employer.

Resumption of Work **69.5** Where the Contractor suspends work or reduces the rate of work, having given notice in accordance with Sub-Clause 69.4, and the Employer subsequently pays the amount due, including interest pursuant to Sub-Clause 60.10, the Contractor's entitlement under Sub-Clause 69.1 shall, if notice of termination has not been given, lapse and the Contractor shall resume normal working as soon as is reasonably possible.

Changes in Cost and Legislation

Increase or Decrease of Cost **70.1** There shall be added to or deducted from the Contract Price such sums in respect of rise or fall in the cost of labour and/or materials or any other matters affecting the cost of the execution of the Works as may be determined in accordance with Part II of these Conditions.

© FIDIC 1987

Subsequent
Legislation

70.2 If, after the date 28 days prior to the latest date for submission of tenders for the Contract there occur in the country in which the Works are being or are to be executed changes to any National or State Statute, Ordinance, Decree or other Law or any regulation or bye-law of any local or other duly constituted authority, or the introduction of any such State Statute, Ordinance, Decree, Law, regulation or bye-law which causes additional or reduced cost to the Contractor, other than under Sub-Clause 70.1, in the execution of the Contract, such additional or reduced cost shall, after due consultation with the Employer and the Contractor, be determined by the Engineer and shall be added to or deducted from the Contract Price and the Engineer shall notify the Contractor accordingly, with a copy to the Employer.

Currency and Rates of Exchange

Currency
Restrictions

71.1 If, after the date 28 days prior to the latest date for submission of tenders for the Contract, the Government or authorised agency of the Government of the country in which the Works are being or are to be executed imposes currency restrictions and/or transfer of currency restrictions in relation to the currency or currencies in which the Contract Price is to be paid, the Employer shall reimburse any loss or damage to the Contractor arising therefrom, without prejudice to the right of the Contractor to exercise any other rights or remedies to which he is entitled in such event.

Rates of
Exchange

72.1 Where the Contract provides for payment in whole or in part to be made to the Contractor in foreign currency or currencies, such payment shall not be subject to variations in the rate or rates of exchange between such specified foreign currency or currencies and the currency of the country in which the Works are to be executed.

Currency
Proportions

72.2 Where the Employer has required the Tender to be expressed in a single currency but with payment to be made in more than one currency and the Contractor has stated the proportions or amounts of other currency or currencies in which he requires payment to be made, the rate or rates of exchange applicable for calculating the payment of such proportions or amounts shall, unless otherwise stated in Part II of these Conditions, be those prevailing, as determined by the Central Bank of the country in which the Works are to be executed, on the date 28 days prior to the latest date for the submission of tenders for the Contract, as has been notified to the Contractor by the Employer prior to the submission of tenders or as provided for in the Tender.

Currencies of
Payment for
Provisional Sums

72.3 Where the Contract provides for payment in more than one currency, the proportions or amounts to be paid in foreign currencies in respect of Provisional Sums shall be determined in accordance with the principles set forth in Sub-Clauses 72.1 and 72.2 as and when these sums are utilised in whole or in part in accordance with the provisions of Clauses 58 and 59.

REFERENCE TO PART II

As stated in the Foreword at the beginning of this document, the FIDIC Conditions comprise both Part I and Part II. Certain Clauses, namely Sub-Clauses 1.1 paragraph (a) (i), 5.1 part 14.1, 14.3, 68.2 and 70.1 must include additional wording in Part II for the Conditions to be complete. Other Clauses may require additional wording to supplement Part I or to cover particular circumstances or the type of work (dredging is an example).

Part II Conditions of Particular Application with guidelines for the preparation of Part II are printed in a separately bound document.

TENDER

NAME OF CONTRACT: * _____

_____

TO: * _____

_____

_____

Gentlemen,

1. Having examined the Conditions of Contract, Specification, Drawings, and Bill of Quantities and Addenda Nos _____ for the execution of the above-named Works, we the undersigned, offer to execute and complete such Works and remedy any defects therein in conformity with the Conditions of Contract, Specification, Drawings, Bill of Quantities and Addenda for the sum of

_____

_____

(_____)

or such other sums as may be ascertained in accordance with the said Conditions.

2. We acknowledge that the Appendix forms part of our Tender.

3. We undertake, if our Tender is accepted, to commence the works as soon as is reasonably possible after the receipt of the Engineer's notice to commence, and to complete the whole of the Works comprised in the Contract within the time stated in the Appendix to Tender.

4. We agree to abide by this Tender for the period of * _____ days from the date fixed for receiving the same and it shall remain binding upon us and may be accepted at any time before the expiration of that period.

5. Unless and until a formal Agreement is prepared and executed this Tender, together with your written acceptance thereof, shall constitute a binding contract between us.

6. We understand that you are not bound to accept the lowest or any tender you may receive.

Dated this _____ day of _____ 19 _____

Signature _____ in the capacity of _____

duly authorised to sign tenders for and on behalf of _____

(IN BLOCK CAPITALS)

Address _____

_____

Witness _____

Address _____

_____

Occupation _____

(Note: All details marked * shall be inserted before issue of Tender documents.)

Appendix

	Sub-Clause	
Amount of security (if any) _____	**10.1**	_____ per cent of the Contract Price
Minimum amount of third party insurance	**23.2**	_____ per occurrence, with the number of occurrences unlimited
Time for issue of notice to commence	**41.1**	_____ days
Time for Completion _____	**43.1**	_____ days
Amount of liquidated damages _____	**47.1**	_____ per day
Limit of liquidated damages _____	**47.1**	_____
Defects Liability Period _____	**49.1**	_____ days
Percentage for adjustment of Provisional Sums	**59.4(c)**	_____ per cent
Percentage of invoice value of listed materials	**60.1(c)**	_____ per cent
Percentage of Retention _____	**60.2**	_____ per cent
Limit of Retention Money _____	**60.2**	_____
Minimum Amount of Interim Certificates __	**60.2**	_____
Rate of interest upon unpaid sums _____	**60.10**	_____ per cent

Initials of Signatory of Tender _____

(Notes: All details in the list above, other than percentage figure against Sub-Clause 59.4, shall be inserted before issue of Tender documents. Where a number of days is to be inserted, it is desirable, for consistency with the Conditions, that the number should be a multiple of seven.

Additional entries are necessary where provision is included in the Contract for:

(a) completion of Sections (Sub-Clauses 43.1 and 48.2(a))
(b) liquidated damages for Sections (Sub-Clause 47.1)
(c) a bonus (Sub-Clause 47.3 — Part II)
(d) payment for materials on Site (Sub-Clause 60.1(c))
(e) payment in foreign currencies (Clause 60 — Part II)
(f) an advance payment (Clause 60 — Part II)
(g) adjustments to the Contract Price on account of Specified Materials (Sub-Clause 70.1 — Part II)
(h) rates of exchange (Sub-Clause 72.2 — Part II))

Agreement

This Agreement made the _____ day of _____ 19 _____

Between _____

of _____

_____ (hereinafter called "the Employer) of the one part and

_____ of _____

(hereinafter called the "Contractor") of the other part

Whereas the Employer is desirous that certain Works should be executed by the contractor, viz _____

and has accepted a Tender by Contractor for the execution and completion of such Works and the remedying of any defects therein

Now this agreement witnesseth as follows:

1. In this Agreement words and expressions shall have the same meanings as are respectively assigned to them in the Conditions of Contract hereinafter referred to.

2. The following documents shall be deemed to form and be read and construed as part of this Agreement, viz:-

   (a) The Letter of Acceptance;
   (b) The said Tender;
   (c) The Conditions of Contract (Parts I and II);
   (d) The Specification;
   (e) The Drawings; and
   (f) The Bill of Quantities.

3. In consideration of the payments to be made by the Employer to the Contractor as hereinafter mentioned the Contractor hereby covenants with the Employer to execute and complete the Works and remedy any defects therein in conformity in all respects with the provisions of the Contract.

4. The Employer hereby covenants to pay the Contractor in consideration of the execution and completion of the Works and the remedying of defects therein the Contract Price or such other sum as may become payable under the provisions of the Contract at the times and in the manner prescribed by the Contract.

   In Witness whereof the parties hereto have caused this Agreement to be executed the day and year first before written in accordance with their respective laws.

   The Common Seal of _____

   _____

   was hereunto affixed in the presence of:-

   or

   Signed Sealed and Delivered by the
   said _____

   in the presence of:

   Binding Signature of Employer _____

   Binding Signature of Contractor _____

CONTENTS

PART II: CONDITIONS OF PARTICULAR APPLICATION

INTRODUCTION

The terms of the Fourth Edition of the Conditions of Contract for Works of Civil Engineering Construction have been prepared by the Fédération Internationale des Ingénieurs Conseils (FIDIC) and are recommended for general use for the purpose of construction of such works where tenders are invited on an international basis. The Conditions are equally suitable for use on domestic contracts.

The version in English of the Conditions is considered by FIDIC as the official and authentic text for the purpose of translation.

The Clauses of general application have been grouped together in a separately bound document and are referred to as Part I — General Conditions. They have been printed in a form which will facilitate their inclusion, as printed, in the contract documents normally prepared.

In the preparation of the Conditions it was recognised that while there are numerous Clauses which will be generally applicable there are some Clauses which must necessarily vary to take account of the circumstances and locality of the Works.

Part I — General Conditions and Part II — Conditions of Particular Application together comprise the Conditions governing the rights and obligations of the parties.

For this reason a Part II standard form has not been produced. It will be necessary to prepare a Part II document for each individual contract and the Guidelines are intended to aid in this task by giving options for various clauses where appropriate.

Part II clauses may arise for one or more reasons, of which the following are examples:

(i) Where the wording in Part I specifically requires that further information is to be included in Part II and the Conditions would not be complete without that information, namely in Sub-Clauses 1.1 paragraphs (a) (i) and (iv), 5.1 (part), 14.1, 14.3, 68.2 and 70.1

(ii) Where the wording in Part I indicates that supplementary information may be included in Part II, but the conditions would still be complete without any such information, namely in Sub-Clauses 2.1 paragraph (b), 5.1 (part), 21.1 paragraph (b) and 72.2.

(iii) Where the type, circumstances or locality of the Works necessitates additional Clauses or Sub-Clauses.

(iv) Where the law of the country or exceptional circumstances necessitate an alteration in Part I. Such alterations should be effected by stating in Part II that a particular Clause, or part of a Clause, in Part I is deleted and giving the substitute Clause, or part, as applicable.

As far as possible, in the Clauses that are mentioned hereunder, example wording is provided. In the case of some Clauses, however, only an aide-memoire for the preparation is given. Before incorporating any example wording it must be checked to ensure that it is wholly suitable for the particular circumstances and, if not, it must be varied. Where example wording is varied and in all cases where additional material is included in Part II, care must be taken that no ambiguity is created with Part I or between the Clauses in Part II.

Dredging and Reclamation Work

Special consideration must be given to Part II where dredging and certain types of reclamation work are involved. Dredgers are considerably more expensive than most items of Contractor's Equipment and the capital value of a dredger can often exceed the value of the Contract on which it is used.

For this reason, it is in the interests of both the Employer and the Contractor that a dredger is operated intensively in the most economic fashion, subject to the quality of work and any other over-riding factors. With this end in view, it is customary to allow the Contractor to execute dredging work continuously by day and by night seven days a week. Another difference from most civil engineering is that on dredging work the Contractor is not normally held responsible for the remedying of defects after the date of completion as certified under Clause 48. Part II contains explanations and example wording to cover the above points and others relating to dredging. Clauses 11, 12, 18, 19, 28, 40, 45, 49, 50 and 51 are those which most often require attention in Part II when dredging work is involved and reference is included under each of these Clauses. Other Clauses may also need additions in Part II in certain circumstances. Reclamation work varies greatly in character and each instance must be considered before deciding whether it is appropriate to introduce in Part II changes similar to those adopted for dredging, or to use the standard civil engineering form unaltered.

PART II CONDITIONS OF PARTICULAR APPLICATION

Clause 1

Sub-Clause 1.1 — Definitions

> *(a) (i) The Employer is (insert name)*

> *(a) (iv) The Engineer is (insert name)*

If further definitions are essential, for example the name of an International Financing Institution (IFI), additions should be made to the list.

Clause 2

Sub-Clause 2.1 — Engineer's Duties

> EXAMPLE

> *The Engineer shall obtain the specific approval of the Employer before carrying out his duties in accordance with the following Clauses of Part I:*

> *(a) Clause (insert applicable number)*

> *(b) Clause (insert applicable number)*

> *(c) Clause (insert applicable number)*

This list should be extended or reduced as necessary.

In some cases the obligation to obtain the approval of the Employer may apply to only one Sub-Clause out of several in a Clause or approval may only be necessary beyond certain limits, monetary or otherwise. Where this is so, the example wording must be varied.

If the obligation to obtain the approval of the Employer could lead to the Engineer being unable to take action in an emergency, where matters of safety are involved, an additional paragraph may be necessary.

> EXAMPLE

> *Notwithstanding the obligation, as set out above, to obtain approval, if, in the opinion of the Engineer, an emergency occurs affecting the safety of life or of the Works or of adjoining property, he may, without relieving the Contractor of any of his duties and responsibilities under the Contract, instruct the Contractor to execute all such work or to do all such things as may, in the opinion of the Engineer, be necessary to abate or reduce the risk. The Contractor shall forthwith comply, despite the absence of approval of the Employer, with any such instruction of the Engineer. The Engineer shall determine an addition to the Contract Price, in respect of such instruction, in accordance with Clause 52 and shall notify the Contractor accordingly, with a copy to the Employer.*

Clause 5

Sub-Clause 5.1 — Language/s and Law

> *(a) The language is (insert as applicable)*
> *(b) The law is that in force in (insert name of country)*

If necessary (a) above should be varied to read:

> *The languages are (insert as applicable)*

and there should be added

> *The Ruling Language is (insert as applicable)*

Sub-Clause 5.2 — Priority of Contract Documents

Where it is decided that an order of precedence of all documents should be included, this Sub-Clause may be varied as follows:

> EXAMPLE
>
> *Delete the documents listed 1 – 6 and substitute:*
>
> *(1) the Contract Agreement (if completed);*
>
> *(2) the Letter of Acceptance;*
>
> *(3) the Tender;*
>
> *(4) the Conditions of Contract Part II;*
>
> *(5) the Conditions of Contract Part I;*
>
> *(6) the Specification;*
>
> *(7) the Drawings; and*
>
> *(8) the priced Bill of Quantities*

or

Where it is decided that no order of precedence of documents should be included, this Sub-Clause may be varied as follows:

> EXAMPLE
>
> *Delete the text of the Sub-Clause and substitute:*
>
> *The several documents forming the Contract are to be taken as mutually explanatory of one another, but in the case of ambiguities or discrepancies the priority shall be that accorded by law. If, in the opinion of the Engineer, such ambiguities or discrepancies make it necessary to issue any instruction to the Contractor in explanation or adjustment, the Engineer shall have authority to issue such instruction.*

Clause 9

Where it is decided that a Contract Agreement should be entered into and executed the form must be annexed to these Conditions.

A suitable form is annexed to Part 1 — General Conditions.

Clause 10

Sub-Clause 10.1 — Performance Security

Where it is decided that a performance security should be obtained by the Contractor, the form must be annexed to these Conditions as stated in Sub-Clause 10.1 of Part I of these Conditions.

Two example forms of performance security are given on pages 7, 8 and 9. The Clause and wording of the example forms may have to be varied to comply with the law of the Contract which may require the forms to be executed under seal.

Where there is provision in the Contract for payments to the Contractor to be made in foreign currency, Sub-Clause 10.1 of Part I of these Conditions may be varied.

> EXAMPLE
>
> *After the first sentence, insert the following sentence:*
>
> *The security shall be denominated in the types and proportions of currencies stated in the Appendix to Tender.*

Where the source of the performance security is to be restricted, an additional Sub-Clause may be added.

EXAMPLE SUB-CLAUSES

Source of Performance Security

10.4 *The performance security, submitted by the Contractor in accordance with Sub-Clause 10.1, shall be furnished by an institution registered in (insert the country where the Works are to be executed) or licensed to do business in such country.*

or

Source of Performance Security

10.4 *Where the performance security is in the form of a bank guarantee, it shall be issued by:*

(a) a bank located in the country of the Employer, or

(b) a foreign bank through a correspondent bank located in the country of the Employer.

Clause 11

Where the bulk or complexity of the data, or reasons of security enforced by the country where the Works are to be executed, makes it impracticable for the Employer to make all data available with the Tender Documents and inspection of some data by the Contractor at an office is therefore expected, it would be advisable to make the circumstances clear.

EXAMPLE SUB-CLAUSE

Access to Data

11.2 *Data made available by the Employer in accordance with Sub-Clause 11.1 shall be deemed to include data listed elsewhere in the Contract as open for inspection at (insert particulars of the office or offices where such data is stored)*

Sub-Clause 11.1 — Inspection of Site

For a Contract comprising dredging and reclamation work the Clause may be varied as follows:

EXAMPLE

In the first paragraph, delete the words 'hydrological and sub-surface' and substitute 'hydrographic and sub-seabed'.

In the second paragraph, under (a) delete the word 'sub-surface' and substitute 'sub-seabed' and under (b) delete the word 'hydrological' and substitute 'hydrographic'.

Clause 12

Sub-Clause 12.2 — Adverse Physical Obstructions or Conditions

For a Contract comprising dredging and some types of reclamation work the Sub-Clause may require to be varied.

EXAMPLE

Delete the words, 'other than climatic conditions on the site,'.

Clause 14

Sub-Clause 14.1 — Programme to be Submitted

The time within which the programme shall be submitted shall be (insert number) days.

Sub-Clause 14.3 — Cash Flow Estimate to be Submitted

The time within which the detailed cash flow estimate shall be submitted shall be (insert number) days.

In both examples given above it is desirable for consistency with the rest of the Conditions that the number of days inserted should be a multiple of seven.

EXAMPLE PERFORMANCE GUARANTEE

By this guarantee We, _____

whose registered office is at _____

(hereinafter called "the Contractor") and _____

whose registered office is at _____

(hereinafter called "the Guarantor") are held and firmly bound unto

_____ *(hereinafter called "the Employer")*

in the sum of _____ *for the payment of which sum*

the Contractor and the Guarantor bind themselves, their successors and assigns jointly and severally by these presents.

Whereas the Contractor by an Agreement made between the Employer of the one part and the Contractor of the other part has entered into a Contract (hereinafter called "the said Contract") to execute and, complete certain Works and remedy any defects therein as therein mentioned in conformity with the provisions of the said Contract.

Now the Condition of the above-written Guarantee is such that if the Contractor shall duly perform and observe all the terms provisions conditions and stipulations of the said Contract on the Contractor's part to be performed and observed according to the true purport intent and meaning thereof or if on default by the Contractor the Guarantor shall satisfy and discharge the damages sustained by the Employer thereby up to the amount of the above-written Guarantee then this obligation shall be null and void but otherwise shall be and remain in full force and effect but no alteration in terms of the said Contract or in the extent or nature of the Works to be executed, completed and defects therein remedied thereunder and no allowance of time by the Employer or the Engineer under the said Contract nor any forbearance or forgiveness in or in respect of any matter or thing concerning the said Contract on the part of the Employer or the said Engineer shall in any way release the Guarantor from any liability under the above-written Guarantee. Provided always that the above obligation of Guarantor to satisfy and discharge the damages sustained by the Employer shall arise only

(a) on written notice from both the Employer and the Contractor that the Employer and the Contractor have mutually agreed that the amount of damages concerned is payable to the Employer or

(b) on receipt by the Guarantor of a legally certified copy of an award issued in arbitration proceeding carried out in conformity with the terms of the said Contract that the amount of the damages is payable to the Employer.

Signed on _____	*Signed on* _____
on behalf of _____	*on behalf of* _____
by _____	*by* _____
in the capacity of _____	*in the capacity of* _____
in the presence of _____	*in the presence of* _____

EXAMPLE SURETY BOND FOR PERFORMANCE

Know all Men by these Presents that (name and address of Contractor)

_____

_____

as Principal (hereinafter called "the Contractor") and (name, legal title and address of Surety) _____

_____

as Surety (herein after called "the Surety"), are held and firmly bound unto (name and address of Employer) _____

_____ *as Obligee (hereinafter called "the Employer") in the amount of* _____ *for the payment of which sum, well and truly to be made, the Contractor and the Surety bind themselves, their successors and assigns, jointly and severally, firmly by these presents.*

Whereas the Contractor has entered into a written contract agreement with the Employer dated the _____ *day of* _____ *19* _____

for (name of Works) _____
in accordance with the plans and specifications and amendments thereto, to the extent herein provided for, are by reference made part hereof and are hereinafter referred to as the Contract.

Now, therefore, the Condition of this Obligation is such that, if the Contractor shall promptly and faithfully perform the said Contract (including any amendments thereto) then this obligation shall be null and void; otherwise it shall remain in full force and effect.

Whenever Contractor shall be, and declared by Employer to be, in default under the Contract, the Employer having performed the Employer's obligations thereunder, the Surety may promptly remedy the default, or shall promptly:-

(1) Complete the Contract in accordance with its terms and conditions; or

(2) Obtain a bid or bids for submission to the Employer for completing the Contract in accordance with its terms and conditions, and upon determination by Employer and Surety of the lowest responsible bidder, arrange for a contract between such bidder and Employer and make available as work progresses (even though there should be a default or a succession of defaults under the contract or contracts of completion arranged under this paragraph) sufficient funds to pay the cost of completion less the balance of the Contract Value; but not exceeding, including other costs and damages for which the Surety may be liable hereunder, the amount set forth in the first paragraph hereof. The term "balance of the Contract Value", as used in this paragraph, shall mean the total amount payable by Employer to Contractor under the Contract, less the amount properly paid by Employer to Contractor; or

(3) Pay the Employer the amount required by Employer to complete the Contract in accordance with its terms and conditions any amount up to a total not exceeding the amount of this Bond.

The Surety shall not be liable for a greater sum than the specified penalty of this Bond.

Any suit under this Bond must be instituted before the issue of the Defects Liability Certificate.

No right of action shall accrue on this Bond to or for the use of any person or corporation other than the Employer named herein or the heirs, executors, administrators or successors of the Employer.

Signed on _____   *Signed on* _____

on behalf of _____   *on behalf of* _____

by _____   *by* _____

in the capacity of _____   *in the capacity of* _____

in the presence of _____   *in the presence of* _____

Clause 15

Where the language in which the Contract documents have been drawn up is not the language of the country in which the Works are to be executed, or where for any other reason it is necessary to stipulate that the Contractor's authorised representative shall be fluent in a particular language, an additional Sub-Clause may be added.

EXAMPLE SUB-CLAUSES

Language Ability of Contractor's Representative

15.2 *The Contractor's authorised representative shall be fluent in (insert name of language).*

or

Interpreter to be Made Available

15.2 *If the Contractor's authorised representative is not, in the opinion of the Engineer, fluent in (insert name of language), the Contractor shall have available on Site at all times a competent interpreter to ensure the proper transmission of instructions and information.*

Clause 16

Where the language in which the Contract documents have been drawn up is not the language of the country in which the Works are to be executed, or where for any other reason it is necessary to stipulate that members of the Contractor's superintending staff shall be fluent in a particular language, an additional Sub-Clause may be added.

EXAMPLE SUB-CLAUSE

Language Ability of Superintending Staff

16.3 *A reasonable proportion of the Contractor's superintending staff shall have a working knowledge of (insert name of language) or the Contractor shall have available on Site at all times a sufficient number of competent interpreters to ensure the proper transmission of instructions and information.*

Where there is a desire, but not a legal requirement, that the Contractor makes reasonable use of materials from or persons resident in the country in which the Works are to be executed, an additional Sub-Clause may be added.

EXAMPLE SUB-CLAUSE

Employment of Local Personnel

16.4 *The Contractor is encouraged, to the extent practicable and reasonable, to employ staff and labour from sources within (insert name of country).*

Clause 18

Sub-Clause 18.1 — Boreholes and Exploratory Excavation

For a Contract comprising dredging and reclamation work the Sub-Clause may require to be varied.

EXAMPLE

Add second sentence as follows:

Such exploratory excavation shall be deemed to include dredging.

Clause 19

Sub-Clause 19.1 — Safety, Security and Protection of the Environment

Where a Contract includes dredging the possibility of pollution should be given particular attention and additional wording may be required. For example, where fishing and recreation areas might be influenced, the Contractor should be required to plan and execute the dredging so that the effect is kept to a minimum. Where there is a risk of chemical pollution from soluble sediments in the dredging area, for instance in a harbour, it is important that sufficient information is provided with the Tender documents. Responsibilities should be clearly defined.

FIDIC 1987

Clause 21

Sub-Clause 21.1 — Insurance of Works and Contractor's Equipment

Where there is provision in the Contract for payments to the Contractor to be made in foreign currency, this Sub-Clause may be varied.

EXAMPLE

Add final sentence as follows:

The insurance in paragraphs (a) and (b) shall provide for compensation to be payable in the types and proportions of currencies required to rectify the loss or damage incurred.

Where it is decided to state the deductible limits for the Employer's Risks, this Sub-Clause may be varied.

EXAMPLE

Add to paragraph (a) as follows:

and with deductible limits for the Employer's Risks not exceeding (insert amounts)

Clauses 21, 23 and 25. Insurances Arranged by Employer

In certain circumstances, such as where a number of separate contractors are employed on a single project, or phased take-over is involved, it may be preferable for the Employer to arrange insurance of the Works, and Third Party insurance. In such case, it must be clear in the Contract that the Contractor is not precluded from taking out any additional insurance, should he desire to do so, over and above that to be arranged by the Employer.

Tenderers must be provided at the Tender stage with details of the insurance to be arranged by the Employer, in order to assess what provision to make in their rates and prices for any additional insurance, and for the amount of policy deductibles which they will be required to bear. Such details shall form part of the Contract between the Employer and the Contractor.

Example wording to allow for the arrangement of insurance by the Employer is as follows:

EXAMPLE

Clause 21

Delete the text of the Clause and substitute the following re-numbered Sub-Clauses:

Insurance of Works 21.1 *Without limiting his or the Contractor's obligations and responsibilities under Clause 20, the Employer will insure:*

(a) the Works, together with materials and Plant for incorporation therein, to the full replacement cost

(b) an additional sum to cover any additional costs of and incidental to the rectification of loss or damage including professional fees and the cost of demolishing and removing any part of the Works and of removing debris of whatsoever nature.

Insurance of Contractor's Equipment 21.2 *The Contractor shall, without limiting his obligations and responsibilities under Clause 20, insure the Contractor's Equipment and other things brought onto the site by the Contractor, for a sum sufficient to provide for their replacement at the Site.*

Scope of Cover 21.3 *The insurance in Sub-Clause 21.1 shall be in the joint names of the Contractor and the Employer and shall cover:*

FIDIC 1987

(a) the Employer and the Contractor against loss or damage as provided in the details of insurance annexed to these Conditions, from the start of work at the Site until the date of issue of the relevant Taking-Over Certificate in respect of the Works or any Section or part thereof as the case may be, and

(b) the Contractor for his liability:
(i) during the Defects Liability Period for loss or damage arising from a cause occurring prior to the commencement of the Defects Liability Period, or

(ii) occasioned by the Contractor in the course of any operations carried out by him for the purpose of complying with his obligations under Clauses 49 and 50

Responsibility for Amounts not Recovered   **21.4**   Any amounts not insured or not recovered from the insurers shall be borne by the Employer or the Contractor in accordance with their responsibilities under Clause 20.

Clause 23

Delete the text of the Clause and substitute:

Third Party Insurance (including Employer's Property)   **23.1**   Without limiting his or the Contractor's obligations and responsibilities under Clause 22, the Employer will insure in the joint names of the Contractor and the Employer, against liabilities for death of or injury to any person (other than as provided in Clause 24) or loss of or damage to any property (other than the Works) arising out of the performance of the Contract, as provided in the details of insurance referred to in Sub-Clause 21.3.

Clause 25

Delete the text of the Clause and substitute:

Evidence and Terms of Insurances   **25.1**   The insurance policies to be arranged by the Employer pursuant to Clauses 21 and 23 shall be consistent with the general terms described in the Tender and copies of such policies shall when required be supplied by the Employer to the Contractor.

Adequacy of Insurances   **25.2**   The Employer shall notify the insurers of changes in the nature, extent or programme for execution of the Works and ensure the adequacy of the insurances at all times in accordance with the terms of the Contract and shall, when required, produce to the Contractor the insurance policies in force and the receipts for payment of the premiums. No variations shall be made to the insurances by the Employer without the prior approval of the Contractor.

Remedy on Employer's Failure to Insure   **25.3**   If and so far as the Employer fails to effect and keep in force any of the insurances referred to in Sub-Clause 25.1, then the Contractor may effect and keep in force any such insurance and pay any premium as may be necessary for that purpose and add the amount so paid to any monies due or to become due to the Contractor, or recover the same as a debt due from the Employer.

Compliance with Policy Conditions   **25.4**   In the event that the Contractor or the Employer fails to comply with conditions imposed by the insurance policies effected pursuant to the Contract, each shall indemnify the other against all losses and claims arising from such failure.

Clause 28

Sub-Clause 28.2 — Royalties

For a Contract comprising dredging and reclamation work and for any other Contract involving the dumping of materials the Sub-Clause may require to be varied.

EXAMPLE

Add second sentence as follows:

The Contractor shall also be liable for all payments or compensation, if any, levied in relation to the dumping of part or all of any such materials.

It is sometimes the case on dredging contracts for the Employer to bear the costs of tonnage and other royalties, rent and other payments or compensation. If such conditions are to apply, Sub-Clause 28.2 should be varied either by adding wording or by deleting the existing wording and substituting new wording.

Clause 31

Where the particular requirements of other contractors are known within reasonable limits at the time of preparation of the Contract documents, details must be stated. The Specification is usually the appropriate place to do so but, exceptionally, some reference may be desirable in the Conditions. In that case, an additional Sub-Clause or Sub-Clauses could be added to this Clause.

Clause 34

It will generally be necessary to add a number of Sub-Clauses, to take account of the circumstances and locality of the Works, covering such matters as: permits and registration of expatriate employees; repatriation to place of recruitment; provision of temporary housing for employees; requirements in respect of accommodation for staff of Employer and Engineer; standards of accommodation to be provided; provision of access roads, hospital, school, power, water, drainage, fire services, refuse collection, communal buildings, shops, telephones; hours and conditions of working; rates of pay; compliance with labour legislation; maintenance of records of safety and health.

EXAMPLE SUB-CLAUSES (to be numbered, as appropriate)

Rates of Wages and Conditions of Labour 34. *The Contractor shall pay rates of wages and observe conditions of labour not less favourable than those established for the trade or industry where the work is carried out. In the absence of any rates of wages or conditions of labour so established, the Contractor shall pay rates of wages and observe conditions of labour which are not less favourable than the general level of wages and conditions observed by other employers whose general circumstances in the trade or industry in which the Contractor is engaged are similar.*

Employment of Persons in the Service of Others 34. *The Contractor shall not recruit or attempt to recruit his staff and labour from amongst persons in the service of the Employer or the Engineer.*

Repatriation of Labour 34. *The Contractor shall be responsible for the return to the place where they were recruited or to their domicile of all such persons as he recruited and employed for the purposes of or in connection with the Contract and shall maintain such persons as are to be so returned in a suitable manner until they shall have left the Site or, in the case of persons who are not nationals of·and have been recruited outside (insert name of country) shall have left (insert name of country).*

Housing for Labour 34. *Save insofar as the Contract otherwise provides, the Contractor shall provide and maintain such accommodation and amenities as he may consider necessary for all his staff and labour, employed for the purposes of or in connection with the Contract, including all fencing, water supply (both for drinking and other purposes), electricity supply, sanitation, cookhouses, fire prevention and fire-fighting equipment, air conditioning, cookers, refrigerators, furniture and other requirements in connection with such accommodation or amenities. On completion of the Contract, unless otherwise agreed with the Employer, the temporary camps/housing provided by the Contractor shall be removed and the site reinstated to its original condition, all to the approval of the Engineer.*

Accident Prevention Officer; Accidents 34. *The Contractor shall have on his staff at the Site an officer dealing only with questions regarding the safety and protection against accidents of all staff and labour. This officer shall be qualified for this work and shall have the authority to issue instructions and shall take protective measures to prevent accidents.*

© FIDIC 1987

Health and Safety	34.	*Due precautions shall be taken by the Contractor, and at his own cost, to ensure the safety of his staff and labour and, in collaboration with and to the requirements of the local health authorities, to ensure that medical staff, first aid equipment and stores, sick bay and suitable ambulance service are available at the camps, housing and on the Site at all times throughout the period of the Contract and that suitable arrangements are made for the prevention of epidemics and for all necessary welfare and hygiene requirements.*
Measures against Insect and Pest Nuisance	34.	*The Contractor shall at all times take the necessary precautions to protect all staff and labour employed on the site from insect nuisance, rats and other pests and reduce the dangers to health and the general nuisance occasioned by the same. The Contractor shall provide his staff and labour with suitable prophylactics for the prevention of malaria and take steps to prevent the formation of stagnant pools of water. He shall comply with all the regulations of the local health authorities in these respects and shall in particular arrange to spray thoroughly with approved insecticide all buildings erected on the Site. Such treatment shall be carried out at least once a year or as instructed by the Engineer. The Contractor shall warn his staff and labour of the dangers of bilharzia and wild animals.*
Epidemics	34.	*In the event of any outbreak of illness of an epidemic nature, the contractor shall comply with and carry out such regulations, orders and requirements as may be made by the Government, or the local medical or sanitary authorities, for the purpose of dealing with and overcoming the same.*
Burial of the Dead	34.	*The Contractor shall make all necessary arrangements for the transport, to any place as required for burial, of any of his expatriate employees or members of their families who may die in (insert name of country). The Contractor shall also be responsible, to the extent required by the local regulations, for making any arrangements with regard to burial of any of his local employees who may die while engaged upon the Works.*
Supply of Foodstuffs	34.	*The Contractor shall arrange for the provision of a sufficient supply of suitable food at reasonable prices for all his staff and labour, or his Subcontractors, for the purposes of or in connection with the Contract.*
Supply of Water	34.	*The Contractor shall, so far as is reasonably practicable, having regard to local conditions, provide on the Site an adequate supply of drinking and other water for the use of his staff and labour.*
Alcoholic Liquor or Drugs	34.	*The contractor shall not, otherwise than in accordance with the Statutes, Ordinances and Government Regulations or Orders for the time being in force, import, sell, give, barter or otherwise dispose of any alcoholic liquor or drugs, or permit or suffer any such importation, sale, gift, barter or disposal by his Subcontractors, agents, staff or labour.*
Arms and Ammunition	34.	*The Contractor shall not give, barter or otherwise dispose of to any person or persons, any arms or ammunition of any kind or permit or suffer the same as aforesaid.*
Festivals and Religious Customs	34.	*The Contractor shall in all dealings with his staff and labour have due regard to all recognised festivals, days of rest and religious or other customs.*
Disorderly Conduct	34.	*The Contractor shall at all times take all reasonable precautions to prevent any unlawful, riotous or disorderly conduct by or amongst his staff and labour and for the preservation of peace and protection of persons and property in the neighbourhood of the Works against the same.*

Clause 35

Additional Sub-Clauses may be desirable to cover circumstances which require the maintenance of particular records or the provision of certain specific reports.

EXAMPLE SUB-CLAUSES (to be numbered, as appropriate)

Records of Safety and Health **35.** The Contractor shall maintain such records and make such reports concerning safety, health and welfare of persons and damage to property as the Engineer may from time to time prescribe.

Reporting of Accidents **35.** The Contractor shall report to the Engineer details of any accident as soon as possible after its occurrence. In the case of any fatality or serious accident, the Contractor shall, in addition, notify the Engineer immediately by the quickest available means.

Clause 40

For a Contract comprising dredging and some types of reclamation work the Clause may be varied.

Sub-Clause 40.1 — Suspension of Work

EXAMPLE

Delete paragraph (c) and renumber paragraph (d) as (c).

Sub-Clause 40.3 — Suspension Lasting more than 84 Days

EXAMPLE

In the first sentence delete the words ', (c) or (d)' and substitute 'or (c)'.

Clause 43

Sub-Clause 43.1 — Time for Completion

Where completion is stated to be by a date and not within a period of time, the Sub-Clause will require to be varied.

EXAMPLE

Delete the words, 'within the time...such extended time' and substitute 'by the date or dates stated in the Appendix to Tender for the whole of the Works or the Section (as the case may be) or such later date or dates'.

Clause 45

For a Contract located in an isolated area, where environmental restrictions do not apply, or where a Contract comprises work, such as dredging and reclamation, that may require continuous working, the Clause may be varied.

EXAMPLE

Delete Sub-Clause 45.1 and substitute:

Working Hours **45.1** *Subject to any provision to the contrary contained in the Contract, the Contractor shall have the option to work continuously by day and by night and on locally recognised days of rest.*

The Contractor's option may be further extended by substituting, in place of the last three words:

holidays or days of rest.

Clause 47

Where it is desired to make provision for the payment of a bonus or bonuses for early completion, an additional Sub-Clause may be added.

In the case where a bonus is provided for early completion of the whole of the Works:

EXAMPLE SUB-CLAUSE

Bonus for
Completion

47.3 *If the Contractor achieves completion of the Works prior to the time prescribed by Clause 43, the Employer shall pay to the Contractor a sum of (insert figure) for every day which shall elapse between the date stated in the Taking-Over Certificate in respect of the Works issued in accordance with Clause 48 and the time prescribed in Clause 43.*

or

In the case where bonuses are provided for early completion of Sections of the Works and details, other than the dates, are given in the Specification:

EXAMPLE SUB-CLAUSE

Bonus for
Completion

47.3 *Sections are required to be completed by the dates given in the Appendix to Tender in order that such Sections may be occupied and used by the Employer in advance of the completion of the whole of the Works.*

Details of the work required to be executed to entitle the Contractor to bonus payments and the amount of the bonuses are stated in the Specification.

For the purposes of calculating bonus payments, the dates given in the Appendix to Tender for completion of Sections are fixed and, unless otherwise agreed, no adjustments of the dates by reason of granting an extension of time pursuant to Clause 44 or any other Clause of these Conditions will be allowed.

Issue of certificates by the Engineer that the Sections were satisfactory and complete by the dates given on the certificates shall, subject to Clause 60, entitle the Contractor to the bonus payments calculated in accordance with the Specification.

Clause 48

Where it can be foreseen that, when the whole of the Works have been substantially completed, the Contractor may be prevented by reasons beyond his control from carrying out the Tests on Completion, an additional Sub-Clause may be added.

EXAMPLE SUB-CLAUSE

Prevention
from Testing

48.5 *If the Contractor is prevented from carrying out the Tests on Completion by a cause for which the Employer or the Engineer or other contractors employed by the Employer are responsible, the Employer shall be deemed to have taken over the Works on the date when the Tests on Completion would have been completed but for such prevention. The Engineer shall issue a Taking-Over Certificate accordingly. Provided always that the Works shall not be deemed to have been taken over if they are not substantially in accordance with the Contract.*

If the Works are taken over under this Sub-Clause the Contractor shall nevertheless carry out the Tests on Completion during the Defects Liability Period. The Engineer shall require the Tests to be carried out by giving 14 days notice.

Any additional costs to which the Contractor may be put, in making the Tests on Completion during the Defects Liability Period, shall be added to the Contract Price.

Clause 49

For a Contract which includes a high proportion of Plant, an additional Sub-Clause may be necessary.

FIDIC 1987

EXAMPLE SUB-CLAUSE

Extension of
Defects Liability

49.5 *The provisions of this Clause shall apply to all replacements or renewals of Plant carried out by the Contractor to remedy defects and damage as if the replacements and renewals had been taken over on the date they were completed. The Defects Liability Period for the Works shall be extended by a period equal to the period during which the Works cannot be used by reason of a defect or damage. If only part of the Works is affected the Defects Liability Period shall be extended only for that part. In neither case shall the Defects Liability Period extend beyond 2 years from the date of taking over.*

When progress in respect of Plant has been suspended under Clause 40, the Contractor's obligations under this Clause shall not apply to any defects occurring more than 3 years after the Time for Completion established on the date of the Letter of Acceptance.

For a Contract comprising dredging work an additional Sub-Clause may be added.

EXAMPLE SUB-CLAUSE

No Remedying
of Defects in
Dredging Work
after Completion

49.5 *Notwithstanding Sub-Clause 49.2, the Contractor shall have no responsibility for the remedying of defects, shrinkages or other faults in respect of dredging work after the date stated in the Taking-Over Certificate.*

Clause 50

For a Contract comprising dredging work and where the second Example Sub-Clause 49.5 has been adopted, an additional Sub-Clause should be added.

EXAMPLE SUB-CLAUSE

No Responsibility for
Cost of Searching of
Dredging Work

50.2 *Notwithstanding Sub-Clause 50.1, the Contractor shall have no responsibility to bear the cost of searching for any defect, shrinkage or other fault in respect of dredging work after the date stated in the Taking-Over Certificate.*

Clause 51

Sub-Clause 51.1 — Variations

For a Contract comprising dredging and some types of reclamation work the Sub-Clause may require to be varied.

EXAMPLE

Add final sentence as follows:

Provided also that the Contractor shall be under no obligation to execute any variation which cannot be executed by the Contractor's Equipment being used or to be used on the Works.

Clause 52

Where provision is made in the Contract for payment in foreign currency, this Clause may be varied.

Sub-Clause 52.1 — Valuation of Variations

EXAMPLE

Add final sentence as follows:

The agreement, fixing or determination of any rates or prices as aforesaid shall include any foreign currency and the proportion thereof.

Sub-Clause 52.2 — Power of Engineer to Fix Rates

Add to first paragraph final sentence as follows:

The agreement or fixing of any rates or prices as aforesaid shall include any foreign currency and the proportion thereof.

Sub-Clause 52.3 — Variations Exceeding 15 per cent

Add final sentence as follows:

The adjustment or fixing of any sum as aforesaid shall have due regard to any foreign currency included in the Effective Contract Price and the proportion thereof.

Where it is required to place some limitation on the range of items for which the rates and prices may be subject to review, the Clause may be varied.

Sub-Clause 52.2 — Power of Engineer to Fix Rates

EXAMPLE

At the end of the first paragraph add:

Provided further that no change in the rate or price for any item contained in the Contract shall be considered unless such item accounts for an amount more than 2 per cent of the Contract Price, and the actual quantity of work executed under the item exceeds or falls short of the quantity set out in the Bill of Quantities by more than 25 per cent.

Clause 54

Where vesting of Contractor's Equipment, Temporary Works and materials in the Employer is required, additional Sub-Clauses may be added.

EXAMPLE WORDING AND SUB-CLAUSES

Sub-Clauses 54.2 and 54.3 shall be renumbered as 54.3 and 54.4 and Sub-Clauses 54.4 to 54.8 shall be renumbered as 54.6 to 54.10. Add additional Sub-Clauses as follows:

Vesting   **54.2**   *All Contractor's Equipment, Temporary Works and materials owned by the Contractor, or by any company in which the Contractor has a controlling interest, shall, when on the Site, be deemed to be the property of the Employer. Provided always that the vesting of such property in the Employer shall not prejudice the right of the Contractor to the sole use of the said Contractor's Equipment, Temporary Works and materials for the purpose of the Works nor shall it affect the Contractor's responsibility to operate and maintain the same under the provisions of the Contract.*

Revesting and   **54.5**   *Upon the removal, with the consent of the Engineer under Sub-Clauses 54.1, of*
Removal   *any such Contractor's Equipment, Temporary Works or materials as have been deemed to have become the property of the Employer under Sub-Clause 54.2, the property therein shall be deemed to revest in the Contractor and, upon completion of the Works, the property in the remainder of such Contractor's Equipment, Temporary Works and materials shall, subject to Clause 63, be deemed to revest in the Contractor.*

Clause 60

Additional Sub-Clauses may be necessary to cover certain other matters relating to payments.

Where payments are to be made in various currencies in predetermined proportions and calculated at fixed rates of exchange the following 3 Sub-Clauses, which should be taken together, may be added:

© FIDIC 1987

EXAMPLE SUB-CLAUSES (to be numbered, as appropriate)

Currency of Account and Rates of Exchange

60. *The currency of account shall be the (insert name of currency) and for the purposes of the Contract conversion between (insert name of currency) and other currencies stated in the Appendix to Tender shall be made in accordance with the Table of Exchange Rates in the Appendix to Tender. Conversion between the currencies stated in such Table other than the (insert name of currency) shall be made at rates of exchange determined by use of the relative rates of exchange between such currencies and the (insert name of currency) set out therein.*

Payments to Contractor

60. *All payments to the Contractor by the Employer shall be made*

(a) in the case of payment(s) under Sub-Clause(s) 70.2 and (insert number of any other applicable Clause), in (insert name of currency/ies);

(b) in the case of payments for certain provisional sum items excluded from the Appendix to Tender, in the currencies and proportions applicable to these items at the time when the Engineer gives instructions for the work covered by these items to be carried out;

(c) in any other case, including Increase or Decrease of Costs under Sub-Clause 70.1, in the currencies and proportions stated in the Appendix to Tender as applicable to such payment provided that the proportions of currencies stated in the Appendix to Tender may from time to time upon the application of either party be varied as may be agreed.

Payments to Employer

60. *All payments to the Employer by the Contractor including payments made by way of deduction or set-off shall be made*

(a) in the case of credit(s) under Sub-Clause(s) 70.2 and (insert number of any other applicable Clause) in (insert name of currency/ies);

(b) in the case of liquidated damages under Clause 47, in (insert name of currency/ies);

(c) in the case of reimbursement of any sum previously expended by the Employer, in the currency in which the sum was expended by the Employer;

(d) in any other case, in such currency as may be agreed.

If the part payable in a particular currency of any sum payable to the Contractor is wholly or partly insufficient to satisfy by way of deduction or set-off a payment due to the Employer in that currency, in accordance with the provisions of this Sub-Clause, then the Employer may if he so desires make such deduction or set-off wholly or partly as the case may be from the balance of such sum payable in other currencies.

Where all payments are to be made in one currency the following Sub-Clause may be added:

EXAMPLE SUB-CLAUSE (to be numbered, as appropriate)

Currency of Account and Payments

60. *The currency of account shall be the (insert name of currency) and all payments made in accordance with the Contract shall be in (insert name of currency). Such (insert name of currency), other than for local costs, shall be fully convertible. The percentage of such payments attributed to local costs shall be as stated in the Appendix to Tender.*

Where place of payment is to be defined the following Sub-Clause may be added:

EXAMPLE SUB-CLAUSE (to be numbered, as appropriate)

Place of Payment

60. *Payments to the Contractor by the Employer shall be made into a bank account nominated by the Contractor in the country of the currency of payment. Where payment is to be made in more than one currency separate bank accounts shall be nominated by the Contractor in the country of each currency and payments shall be made by the Employer accordingly.*

Where provision is to be included for an advance payment the following Sub-Clause may be added:

EXAMPLE SUB-CLAUSE (to be numbered, as appropriate)

Advance Payment **60.** *An advance payment of the amount stated in the Appendix to Tender shall, following the presentation by the Contractor to the Employer of an approved performance security in accordance with Sub-Clause 10.1 and a Guarantee in terms approved by the Employer for the full value of the advance payment, be certified by the Engineer for payment to the Contractor. Such Guarantee shall be progressively reduced by the amount repaid by the Contractor as indicated in interim certificates of the Engineer issued in accordance with this Clause. The advance payment shall not be subject to retention. The advance payment shall be repaid by way of reduction in interim certificates commencing with the next certificate issued after the total certified value of the Permanent Works and any other items in the Bill of Quantities (excluding the deduction of retention) exceeds (insert figure) per cent of the sum stated in the Letter of Acceptance. The amount of the reduction in each interim certificate shall be one (insert fraction) of the difference between the total value of the Permanent Works and any other items in the Bill of Quantities (excluding the deduction of retention) due for certification in such interim certificate and the said value in the last preceding interim certificate until the advance payment has been repaid in full. Provided that upon the issue of a Taking-Over Certificate for the whole of the Works or upon the happening of any of the events specified in Sub-Clause 63.1 or termination under Clauses 65, 66 or 69, the whole of the balance then outstanding shall immediately become due and payable by the Contractor to the Employer.*

Clause 67

Where it is decided that a settlement of dispute procedure, other than that of the International Chamber of Commerce (ICC), should be used the Clause may be varied.

Sub-Clause 67.3 — Arbitration

EXAMPLE

Following paragraph (b), delete the words 'shall be finally settled...International Chamber of Commerce' and substitute 'shall be finally settled under the UNCITRAL Arbitration Rules as administered by (insert name of administering authority)'.

Where alternatives to ICC are considered care should be taken to establish that the favoured alternative is appropriate for the circumstances of the Contract and that the wording of Clause 67 is checked and amended as may be necessary to avoid any ambiguity with the alternative.

Clause 68

Sub-Clause 68.2 — Notice to Employer and Engineer

For the purposes of this Sub-Clause the respective addresses are:

(a) The Employer (insert address)

(b) The Engineer (insert address)

The addresses should be inserted when the documents are being prepared prior to inviting tenders.

Clause 69

Sub-Clause 69.1 — Default of Employer

Where the Employer is a government it may be considered appropriate to vary the Sub-Clause.

© FIDIC 1987

EXAMPLE

Delete paragraph (c) and renumber paragraph (d) as (c).

Where the terms of the Sub-Clause, when read in conjunction with Sub-Clause 69.3, are in conflict with the law of the country the Sub-Clause may require to be varied.

EXAMPLE

Delete 'or' at the end of paragraph (c) and delete paragraph (d).

Clause 70

Three alternative methods of dealing with price adjustment are given below.

The first alternative is suitable where a contract is of short duration and no price adjustment is to be made:

Sub-Clause 70.1 — Increase or Decrease in Cost

EXAMPLE

Delete the text of the Sub-Clause and substitute

Subject to Sub-Clause 70.2, the Contract Price shall not be subject to any adjustment in respect of rise or fall in the cost of labour, materials or any other matters affecting the cost of execution of the Contract.

Sub-Clause 70.2 — Subsequent Legislation

EXAMPLE

Delete the words, 'other than under Sub-Clause 70.1,'.

The second alternative is suitable where price adjustment is to be made by establishing the difference in cost between the basic price and the current price of local labour and specified materials:

Sub-Clause 70.1 — Increase or Decrease in Cost

EXAMPLE

Delete the text of the Sub-Clause and substitute

Adjustments to the Contract Price shall be made in respect of rise or fall in the cost of local labour and specified materials as set out in this Sub-Clause.

(a) Local Workmen

   (i) For the purpose of this Sub-Clause:

"Local Workmen" means skilled, semi-skilled and unskilled workmen of all trades engaged by the Contractor on the Site for the purpose of or in connection with the Contract or engaged full time by the Contractor off the Site for the purpose of or in connection with the Contract (by way of illustration but not limitation: workmen engaged full time in any office, store, workshop or quarry).

"Basic Rate" means the applicable basic minimum wage rate prevailing on the date 28 days prior to the latest date for submission of tenders by reason of any National or State Statute, Ordinance, Decree or other Law or any regulations or bye-law of any local or other duly constituted authority, or in order to conform with practice amongst good employers generally in the area where the Works are to be carried out.

"Current Rate" means the applicable basic minimum wage rate for Local Workmen prevailing on any date subsequent to the date 28 days prior to the latest date set for submission of tenders by reason of any National or State Statute, Ordinance, Decree or other Law or any regulation or bye-law of any local or other duly constituted authority, or in order to conform with practice amongst good employers generally in the area where the works are to be carried out.

(ii) The adjustment to the Contract Price under the terms of this Sub-Clause shall be calculated by multiplying the difference between the Basic and Current Rates for Local workmen by:

(a) the number of all hours actually worked, and also

(b) in respect of those hours worked at overtime rates, by the product of the number of said hours and the percentage addition required by the law to be paid by the Contractor for overtime.

Such adjustment may be either an addition to or a deduction from the Contract Price.

(iii) No other adjustment of the Contract Price on account of fluctuation in the remuneration of Local Workmen shall be made.

(b) Specified Materials

(i) For the purpose of this Sub-Clause:

"Specified Materials" means the materials stated in Appendix (insert reference) to Tender required on the Site for the execution and completion of the Works.

"Basic Prices" means the current prices for the specified materials prevailing on the date 28 days prior to the latest date for submission of tenders.

"Current Prices" means the current prices for the specified materials prevailing at any date subsequent to the date 28 days prior to the latest date for submission of tenders.

(ii) The adjustment to the Contract Price under the terms of this Sub-Clause shall be calculated by applying the difference between the Basic and Current Prices to the quantity of the appropriate Specified Material which is delivered to the Site during the period for which the particular Current Price is effective. Such adjustment may be either an addition to or a deduction from the Contract Price.

(iii) The Contractor shall use due diligence to ensure that excessive wastage of the Specified Materials shall not occur. Any Specified Materials removed from the Site shall be clearly identified in the records required under paragraph (d) of this Sub-Clause.

(iv) The provisions of this Sub-Clause shall apply to fuels used in Contractor's Equipment engaged on the Site for the purposes of executing the Works, including vehicles owned by the Contractor (or hired by him under long term arrangements under which the Contractor is obligated to supply fuel) engaged in transporting any staff, labour, Contractor's Equipment, Temporary Works, Plant or materials to and from the Site. Such fuels shall be clearly identified in the records required under paragraph (d) of this Sub-Clause. The provisions of this Sub-Clause shall not apply to any fuels sold or supplied to any employee of the Contractor or to any person for use in any motor vehicle not being used for the purposes of the Contract.

(v) The Contractor shall at all times have regard to suitable markets and shall, whenever buying materials a variation in the cost of which would give rise to an adjustment of the Contract Price under this Sub-Clause, be diligent to buy or procure the same at the most economical prices as are consistent with the due performance by the Contractor of his obligations under the Contract.

If at any time there shall have been any lack of diligence, default or negligence on the part of the Contractor, whether in observing the above requirements or otherwise, then, for the purposes of adjusting the Contract Price pursuant hereto, no account shall be taken of any increase in cost which may be attributable to such lack of diligence, default or negligence and the amount by which any cost would have been decreased but for such lack of diligence, default or negligence shall be deducted from the Contract Price.

(vi) No other adjustment to the Contract Price on account of fluctuation in the cost of materials shall be made.

(c) Overheads and Profits Excluded

In determining the amount of any adjustment to the Contract Price pursuant to this Sub-Clause no account shall be taken of any overheads or profits.

(d) Notices and Records

The Contractor shall forthwith, upon the happening of any event which may or may be likely to give rise to adjustment of the Contract Price pursuant to this Sub-Clause, give notice thereof to the Engineer and the Contractor shall keep such books, accounts and other documents and records as are necessary to enable adjustment under this Sub-Clause to be made and shall, at the request of the Engineer, furnish any invoices, accounts, documents or records so kept and such other information as the Engineer may require.

(e) Adjustment after Date of Completion

Adjustment to the Contract Price, after the due date for completion of the whole of the Works pursuant to Clause 43, or after the date of completion of the whole of the Works certified pursuant to Clause 48, shall be made in accordance with Current Rates or Current Prices, as applicable, ruling at the due date for completion or the date stated in the Taking-Over Certificate, whichever is the earlier.

(f) Determination of Adjustment to Contract Price

The amount of any adjustment to the Contract Price pursuant to this Sub-Clause shall be determined by the Engineer in accordance with the foregoing rules.

EXAMPLE APPENDIX TO TENDER
for use in conjunction with the second alternative.

SPECIFIED MATERIALS

MATERIAL	UNIT	PRICE AND LOCATION	TRANSPORT TO SITE	PRICE DELIVERED TO SITE	REMARKS
Bitumen					
Diesel Petrol Lubricants					
Cement					
Reinforcing Steel					
Explosives					

NOTES:

1. The Contractor shall provide copies of quotations to substantiate all prices included in the above table.

2. All subsequent price substantiation shall be from the same source as original unless otherwise agreed by the Engineer.

3. The Contractor shall submit full explanation and provide substantiating documentation for the mode of transport to Site he proposes. Only the proposed documented mode of transport shall qualify for price adjustment.

(Note: Materials stated in the Appendix to Tender should be those of which substantial quantities are involved.)

FIDIC 1987

The third alternative is suitable where price adjustment is to be made through the application of indices in a formula:

Sub-Clause 70.1 — Increase or Decrease in Cost

EXAMPLE

Delete the text of the Sub-Clause and substitute

(a) Adjustments to the Contract Price in respect of rise and fall in the cost of labour and materials and other matters affecting the cost of execution of the Works shall be calculated for each monthly statement pursuant to Sub-Clause 60.1, the Statement at Completion pursuant to Sub-Clause 60.5 and the Final Statement pursuant to Sub-Clause 60.6 in accordance with the provisions of this Sub-Clause if there shall be any changes in the following Index figures compiled by (insert details of source of indices) and published by (insert details of publication):-

(i) the Index of the cost of Labour in (insert name of country)

(ii) the Index of the cost of (insert other factor, as relevant)

(iii) the Index of the cost of (insert other factor, as relevant)

(b) For the purpose of this Sub-Clause:

(i) "Base Index Figure" shall mean the index figure applicable on the date 28 days prior to the latest date for submission of tenders.

(ii) "Current Index Figure" shall mean the index figure applicable on the last day of the period to which the particular statement relates.

Provided that in respect of any work the value of which is included in any such monthly statement (or Statement at Completion or Final Statement) and which was executed after the due date (or extended date) for completion of the whole of the Works, pursuant to Clause 43, the Current Index Figure shall be the index figure applicable on the aforesaid due date (or extended date) for completion of the whole of the Works.

(iii) "Effective Value" shall be the difference between:

(a) The amount which is due to the Contractor under the provisions of Sub-Clauses 60.2, 60.5 or 60.8 (before deducting retention and excluding repayment of the advance payment) less any amounts for:

work executed under nominated Subcontracts

materials and Plant on the Site, as referred to in Sub-Clause 60.1 (c)

dayworks, variations or any other items based on actual cost or current prices, and bonuses (if any)

adjustments under Clause 70,

and

(b) The amount calculated in accordance with (b) (iii) (a) of this Sub-Clause and included in the last preceding statement.

(c) The adjustment to the Contract Price shall be calculated by multiplying the Effective Value by a Price Fluctuation Factor which shall be the net sum of the products obtained by multiplying each of the proportions given in paragraph (d) of this Sub-Clause by the following fraction:

$$\frac{\text{Current Index Figure} - \text{Based Index Figure}}{\text{Base Index Figure}}$$

calculated using the relevant index figures.

(d) For the purpose of calculating the Price Fluctuation Factor, the proportions referred to in paragraph (c) of this Sub-Clause shall (irrespective of the actual constituents of the work) be as follows:

0.   *in respect of labour (and supervision) costs subject to adjustment by reference to the Index referred to in (a) (i) of this Sub-Clause;*

0.   *in respect of   by reference to the Index referred to in (a) (ii) of this Sub-Clause;*

0.   *in respect of   by reference to the Index referred to in (a) (iii) of this Sub-Clause;*

0.   *in respect of all other costs which shall not be subject to any adjustment;*

1.00 *Total*

(e) Where the value of an Index is not known at the time of calculation, the latest available value shall be used and any adjustment necessary shall be made in subsequent monthly statements.

(Note: The number of indices included under (a) of this Sub-Clause may be varied, if it is determined that a different number of factors should be separately identified, and in such case (d) of this Sub-Clause must be altered to be consistent.)

Clause 72

Sub-Clause 72.2 — Currency Proportions

Where it is decided that the rate or rates of exchange shall be established from a source other than the Central Bank of the country, the Sub-Clause may be varied.

EXAMPLE

Delete the words from 'prevailing...' to the end of the sentence and substitute

'stated in the Appendix to Tender'.

Clause 73 onwards

Where circumstances require, additional Clauses may be added.

EXAMPLE CLAUSES (to be numbered, starting with Clause 73, as appropriate).

Where the law applicable to the Contract does not cover bribery, the following example Clause may be added.

Bribes   *.1 If the Contractor or any of his Sub-contractors, agents or servants offers to give or agrees to offer or give to any person, any bribe, gift, gratuity or commission as an inducement or reward for doing or forbearing to do any action in relation to the Contract or any other contract with the Employer or for showing or forbearing to show favour or disfavour to any person in relation to the Contract or any other contract with the Employer, then the Employer may enter upon the Site and the Works and terminate the employment of the Contractor and the provisions of Clause 63 hereof shall apply as if such entry and termination had been made pursuant to that Clause.*

Where circumstances require that particular confidentiality is observed, the following example Clause may be added.

Details to be Confidential

.1 *The Contractor shall treat the details of the Contract as private and confidential, save in so far as may be necessary for the purposes thereof, and shall not publish or disclose the same or any particulars thereof in any trade or technical paper or elsewhere without the previous consent in writing of the Employer or the Engineer. If any dispute arises as to the necessity of any publication or disclosure for the purpose of the Contract the same shall be referred to the decision of the Employer whose award shall be final.*

Where the Contract is being financed wholly or in part by an International Financing Institution whose Articles require a restriction on the use of the funds provided, the following example Clause may be added.

Expenditure Restricted

.1 *The Contractor shall not make any expenditures for the purpose of the Contract in the territories of any country which is not a member of (insert name of International Financing Institution) not shall he make any expenditure for goods produced in or services supplied from such territories.*

Where the Contractor may be a joint venture, the following example Clause may be added.

Joint and Several Liability

.1 *If the Contractor is a joint venture of two or more persons, all such persons shall be jointly and severally bound to the Employer for the fulfilment of the terms of the Contract and shall designate one of such persons to act as leader with authority to bind the joint venture. The composition or the constitution of the joint venture shall not be altered without the prior consent of the Employer.*

Index

References to Clause numbers in the FIDIC Conditions (appendix 2) are in italic type. An asterisk after the Clause number signifies that the Clause is mentioned in Part 2 of the FIDIC Conditions. References to page numbers in the main text of the Digest are in roman type.